Analytical Troubleshooting of Process Machinery and Pressure Vessels: Including Real-World Case Studies

Analytical Troubleshooting of Process Machinery and Pressure Vessels: Including Real-World Case Studies

Editor

Chetna Saraswat

scitus
academics

Analytical Troubleshooting of Process Machinery and Pressure Vessels: Including Real-World Case Studies

Edited by **Chetna Saraswat**

Printed in 2017

ISBN: 978-1-68117-333-7

Library of Congress Control Number: 2015939245

© 2016 by
SCITUS Academics LLC,
616, Corporate Way, Suite 2, 4766,
Valley Cottage, NY 10989

www.scitusacademics.com

Contents

vi

Preface

A highly practical troubleshooting tool for today's complex processing industry, Evolving industrial technology-driven by the need to increase safety while reducing production losses-along with environmental factors and legal concerns has resulted in an increased emphasis on sound troubleshooting techniques and documentation. Analytical Troubleshooting of Process Machinery and Pressure Vessels provides both students and engineering professionals with the tools necessary for understanding and solving equipment problems in today's complex processing environment. It is a practical book that has been used to design and troubleshoot over 90% of the equipment worked. A rough estimate is that the examples in this book have saved over $50 million in lost production or warranty claims by eliminating repeat failures or by avoiding failures altogether.

Editor

Ultra-Rapid Elimination of Biofilms via the Combustion of a Nanoenergetic Coating

Byung-Doo Lee[1], Rajagopalan Thiruvengadathan[2, 3], Sachidevi Puttaswamy[1], Brandon M Smith[1], Keshab Gangopadhyay[2, 3], Shubhra Gangopadhyay[2, 3], and Shramik Sengupta[1]

[1]Department of Biological Engineering, University of Missouri, 1406 E Rollins St., 252 Ag Engineering Building, Columbia, MO 65211-5200, USA

[2]Department of Electrical and Computer Engineering, University of Missouri, 349, Engineering Building West, Columbia, MO 65211, USA

[3]NEMS/MEMS Works LLC, 8850 Westlake Road West, Columbia, MO 65202, USA

ABSTRACT

Background

Biofilms occur on a wide variety of surfaces including metals, ceramics, glass etc. and often leads to accumulation of large number of various microorganisms on the surfaces. This biofilm growth is highly undesirable in most cases as biofilms can cause degradation of the instruments and its performance along with contamination of the samples being processed in those systems. The current "offline" biofilm removal methods are effective but labor intensive and generates waste streams that are toxic to be directly disposed. We present here a novel process that uses nano-energetic materials to eliminate biofilms in < 1 second. The process involves spray-coating a thin layer of nano-energetic material on top of the biofilm, allowing it to dry, and igniting the dried coating to incinerate the biofilm.

Results

The nanoenergetic material is a mixture of aluminum (Al) nanoparticles dispersed in a THV-220A (fluoropolymer oxidizer) matrix. Upon ignition, the Al nanoparticles react with THV-220A exothermically, producing high temperatures (>2500 K) for an extremely brief period (~100 ms) that destroys the biofilm underneath. However, since the total amount of heat produced is low (~0.1 kJ/cm^2), the underlying surface remains undamaged. Surfaces with biofilms ofPseudomonas aeruginosa initially harboring ~ 10^7 CFU of bacteria /cm^2 displayed final counts of less than 5 CFU/cm^2 after being subjected to our process. The byproducts of the process consist only of washable carbonaceous residue and gases, making this process potentially inexpensive due to low toxic-waste disposal costs.

Conclusions

This novel method of biofilm removal is currently in the early stage of development. However, it has potential to be used in offline biofilm elimination as a rapid, easy and environmentally friendly method.

BACKGROUND

A biofilm is defined as a microbially derived sessile community characterized by cells that are irreversibly attached to a substratum or interface or to each other; are embedded in a matrix of extracellular polymeric substances that they have produced; and exhibit an altered phenotype with respect to growth rate and gene transcription [1]. Biofilms can occur spontaneously (without deliberate intention to grow them) on a wide variety of surfaces such as metals, plastics, glass, ceramics, wood and cement. Once established, they can accommodate a large number of bacteria per unit area of the surface. While $\sim 10^5$ - 10^7 CFU (Colony Forming Units) of bacteria /cm^2 are commonly encountered, numbers as high as $10^9 - 10^{10}$ CFU/cm^2 have been reported [2,3].

Their presence may be undesirable in a variety of applications. For instance, on ship hulls, the formation of a microbial biofilm can raise the drag coefficient by as much as 29% [4], contributing to correspondingly higher fuel usage. In heat exchangers and cooling water systems, which are an integral part of a wide variety of industrial processes, a 250 micron thick layer of biofilm may reduce the effective heat transfer coefficient of a heat exchanger by as much as 50% [5]. In addition, the metabolism of bacteria in the biofilm (production of carbonic, pyruvic, citric, lactic and other acids) causes a reduction in pH at the surfaces, leading to enhanced rate of chemical corrosion [6]. This inflicts additional economic burdens such as the need for premature replacement of equipment and unscheduled downtime to clean fouled equipment [7]. In the oil and natural gas industry, bacterial biofilms cause financial losses of $\sim$ $100 Million each year through the corrosion of pipelines and process equipment and souring of reservoirs [8]. In the paper manufacturing industry, biofilms are responsible for an estimated 10-20% of all machine downtime[9]. Thus, there are huge incentives to (a) prevent biofilm formation, and (b) to minimize their growth rate during operation of a wide variety of process equipments like tanks, transport tubing, and heat exchangers. Consequently, several approaches have been explored in the past. These approaches include the use of materials and coatings that hinder biofilm formation and growth, the continuous or pulsed addition of chemicals such as acids, oxidizers or enzymes to the process fluid, and the intermittent use of mechanical cleaning agents like scrubbing balls. Despite these efforts, it is almost impossible to completely prevent biofilms from getting established, and as a result, adversely affecting the performance of the equipment [10]. Once the performance of the equipment falls below acceptable levels, they have been taken offline for biofilm removal.

The offline removal of biofilms from process equipment is also a difficult task. The common methods adopted for the offline removal of biofilm from process equipment [10] can be broadly classified into mechanical and chemical processes. The most common mechanical processes include water/steam/sand blasting for large exposed surfaces (blasting being the process of forcibly propelling a stream of material against a surface under high pressure) and abrasive pads for smaller, more difficult to reach surfaces such as the interior of tubes. The main disadvantages of using mechanical processes are that they are labor intensive and take a long time. The latter is especially undesirable, as in many cases, the whole process remains shut for the duration during which one or more of the equipments are brought offline, resulting in losses of tens of thousands of dollars an hour. The other alternative is to use strong chemical cleaning agents like acids, alkalis, and strong biocides. Strong chemicals are often required because the biofilm's extracellular matrix prevents milder chemicals such as antibiotics and germicides from acting on the cells embedded within it. The use of chemicals for biofilm removal has its own advantages and disadvantages. While they are usually less labor intensive, relatively faster, and can act on hard-to-reach surfaces, they are often expensive. Moreover, the use of strong chemicals can also result in the generation of waste-streams that are expensive to dispose off due to their toxicity.

Thus, there is a need for an offline biofilm-removal process for process equipment that is fast, effective, economical, and yet environmentally friendly. Table 1 lists numerous approaches (ultrasonication, electric fields, mild chemicals such as enzymes, and their combinations [11-17]) that have been employed and reported by other groups for this purpose. As can be seen, their efficacy is limited (they achieve only 1 to 3 $\log_{10}$ reductions in the number of viable bacteria per unit area) and/or take a long time (hours). In contrast, if our proposed method were employed for the same application (offline removal of biofilms from process equipment), the biofilm removal could be potentially completed faster (in minutes), and with greater efficacy (> 5 $\log_{10}$ reduction in the number of viable bacteria).

Table 1: Efficacy of various environmentally friendly processes used for the removal of biofilms

Biofilm type	Method used for biofilm elimination	Log reduction in CFU count	Time taken
P. aeruginosa and S. aureus	Cleaning with detergents, followed by high-pressure wash and mechanical scrubbing [11]	< 3	20 min soak detergent + 1 min wash + <1 min scrub
P. aeruginosa and K. pneumoniae	Treatment with multiple chemicals (chelating agents, hypochlorites etc.) [12]	1–3 depending on chemical used	1 hr
Pseudomonas fluorescens	Combination of Enzymes (proteolytic + polysaccharide-degrading enzymes) [13]	2-4	2 hrs
P. aeruginosa, S. epidermidis, and S. aureus	Ultrasound [15]	1.02 – 1.48	10 minutes
E. coli and S. aureus	Chelating Agents (EDTA / EGTA) and Ultrasound [14]	~ 2	Unknown soak time; 10–60s sonication
Pseudomonas aeruginosa	Biocides (Chemicals) + Electric Field [16]	3	12 hours
E. coli	High Pressure CO_2 / N_2 aerosols [17]	1-2	90 s
P. aeruginosa	Our Method (Rapid combustion of a sprayed on layer of nano-energetic materials)	> 5	~ 1 min spray; < 1 s burn

Lee et al.

Lee et al. BMC Biotechnology 2013 13:30   doi:10.1186/1472-6750-13-30.

We present here a novel material and method that is able to outperform the methods listed in Table 1 in both speed and efficacy. Briefly, we use an optimized blend of Al nanoparticles (fuel) in a fluoropolymer (oxidizer) matrix that is spray-coated onto the surfaces, and which burns away extremely rapidly (< 1 sec/cm^2), generating very high temperatures [2200–3200 K [18]] that destroys the biofilm, but leaves the underlying surface intact. The underlying surface remains unaffected because the amount of heat released is not very high: ~ 0.1 kJ/cm^2, according to our estimates based on the heat of combustion of Al [19], and the known loads of Al nanoparticles in our formulation. The key to the efficacy of the process lies in the use of Al nanoparticles with average size of 80 nm and narrow size distribution. The nanometer size of Al particles not only enables the rapid release of the heat of combustion (significantly reduced mass transfer limitation) along with the generation of high temperatures, but also allows us to spread a small mass of Al (~ 10 mg) uniformly over the test areas (~20 cm^2). The latter limits the amount of heat released, which, in turn, limits the damage done to the underlying substrate. We demonstrate the efficacy of this technique using Pseudomonas aeruginosa biofilms grown at moderate shear as our model biofilm. P. aeruginosa was chosen because it has been extensively studied [20], and to compare our technique to those of other researchers [11-13] who report the efficacy of their biofilm removal/killing methods using P. aeruginosa biofilms.

METHODS

Cultivation of Model Biofilms on Substrates of Interest

We cultivated P. aeruginosa biofilms on a variety of substrates such as metals (steel and brass), ceramics (bathroom tiles), and glass that can be expected to withstand the high temperature generated during the burning process for a very short duration.

Above-mentioned substrates with dimensions of 1" × 3" served as our test coupons (except for the ceramic, for which a 2" × 2" piece was used instead). These substrates were first thoroughly cleaned to ensure no pre-existing biofilm. A 1% (w/v) solution of detergent was prepared, and the substrates were first cleaned by sonicating them in this solution for 10 minutes using an ultrasonic bath. The detergent solution was then replaced with DI water and the substrates were sonicated for an additional 10 minutes. The substrates were then rinsed with DI water and then placed in 2 M HCl (for glass and ceramic), or bleach solution (for metals). The substrates were then sonicated again in DI water and rinsed. Finally, they were air-dried in a Biological Safety Cabinet.

Cultures of P. aeruginosa were obtained from commercial sources (Ward's Natural Sciences), and an aliquot was inoculated into 10 ml of TSB (Tryptic Soy Broth) and incubated overnight with shaking at 37°C. The resulting log-culture had a concentration of ~10^9 CFU/ml. The bacterial cells were isolated by centrifugation, and re-suspended in an equal volume (10 ml) of 1× Phosphate Buffered Saline (PBS) (a buffer consisting of Sodium Chloride and Sodium Phosphate). This suspension of bacteria was then introduced into a sterile (autoclaved) vessel with a capacity of ~1 L loaded with ~500 ml of 1/10 × TSB. Multiple (four to six) coupons of a particular material (metal, glass or ceramic) were prepared by covering one side with a piece of adhesive backed silicone rubber sheet, loaded into the tank with the exposed side upwards (in contact with the liquid), the top of the tank covered in saran wrap, and the tank placed in an incubator-shaker. The incubator shaker was operated at room temperature (~ 25°C) with an oscillation speed of 200 rpm, which corresponds to a shear rate of ~10^5 s^{-1} at the fluid-solid interface (biofilm). The biofilm was allowed to form over a period of 4 days. At the end of this period, during which there was perceptible growth of biofilm in the system, the coupons were extracted, washed in DI water to remove cells that adhere weakly to the surface (those not within the biofilm matrix) and loaded into individual Ziploc™ bags (pre-sterilized by wiping with 70% ethanol and exposed to UV radiation in a biological safety cabinet) and stored in refrigerator (4°C). They were then used (within a period of 3 days) for further testing. It may be noted that the biofilms still retain their characteristic slimy appearance after retrieval from storage, indicating that they remain in a hydrated state.

The Biofilm Removal Process

Our proposed process to eliminate biofilms from substrates of interest is illustrated schematically in Figure 1. As shown in Figure 1, we begin the process by dissolving a known amount of THV 220A in acetone using sonication. THV 220A is a commercially available (3 M, St. Paul, MN) fluoropolymer, composed of tetrafluoroethylene, hexafluoropropylene and vinylidene-fluoride. Al nanoparticles are then added to this solution and dispersed homogeneously using an ultrasonic bath. The fluoropolymer plays the role of oxidizer and the Al nanoparticles plays the role of fuel in the nanoenergetic composition. The amount of acetone used in the dispersions was varied as 7, 8, and 10 ml for a total mass of 500 mg of THV 220A polymer and Al nanoparticles. The amount of Al nanoparticles and THV 220A were varied suitably so that the weight ratio of Al to THV 220A was kept at 1:9, 2:8, and 3:7. The nanoenergetic dispersion was then sprayed uniformly on top of the surfaces of interest (biofilm-covered surfaces of different materials).The acetone evaporates rapidly, leaving behind a dry, paint-like coating on the surface. In order to keep the thickness of the coating nearly the same on any given substrate, the same volume (7 ml) of fluoropolymer/nanoparticle and acetone blend was uniformly used, yielding nanoenergetic coatings 80–100 microns thick. The dried layer was then ignited at one corner using a small flame-torch. The flame self-propagated extremely rapidly, and consumed the whole coated surface (~ 3 inch × 1 inch), after which it exhausted itself. The whole process (initiation, propagation, and quenching /exhaustion) took less than 1 second for the surfaces tested (1 inch × 3 inch pieces), and left behind a dark, flaky residue, which could be blown away and/or rinsed off to obtain the clean, biofilm-free surface underneath. Based on our earlier studies of the similar blends for other applications [21], the residue is believed to be carbonaceous, with minor amounts of aluminum oxide. (Most of the aluminum is oxidized to AlF_3 by the fluoropolymer). The amount of acetone and the weight ratio of Al nanoparticles to THV 220A were optimized by observing how well the flame self-propagated upon ignition throughout the surface. More importantly, during this optimization, it was ensured that the swiftly propagated flame only destroyed the biofilm, while not significantly damaging the substrate underneath.

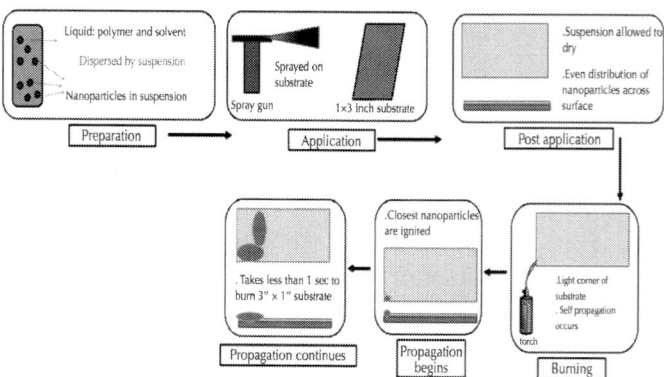

Figure 1: Represents the schematic of the proposed process for ultra-rapid removal of biofilms, which involves spray coating the nanoparticles in suspension onto the substrate containing biofilm, followed by burning the entire surface of the substrate by initiating the ignition of one of its corners.

Upon combustion, nano-aluminum is known to produce temperatures of 2200 K–3200 K [18]. Such high temperatures are likely to destroy biofilms and organisms harbored within. Because the duration of the temperature pulse is small, (less than 10 ms at any point on the substrate) and the amount of heat imparted is relatively low (0.1 kJ/cm²), materials such as metals and ceramics that have high thermal diffusivity, are unaffected by the process.

Assessment of Bacterial Numbers in Biofilms (Before and After Combustion)

To estimate the surface density (number of bacteria per unit area) of bacteria present in the biofilm that grew on a coupon, we performed the following procedures: First, using an autoclaved razor, we shaved off the biofilm from the surface of interest. The shavings were collected in a plastic centrifuge tube with 7 ml of sterile phosphate buffered saline (1× PBS) solution. The surface and razor blade were also rinsed with PBS after shaving, and the rinsed solution pooled with the solution into which the shavings were deposited. The volume was then made up to 10 ml. In order to disperse the bacteria lodged in the peptidoglycan matrix, the sample was vortexed and then sonicated at low power in a Branson 2510 sonicator for 2 minutes. An estimate of the total number of bacteria present in this 10 ml volume was obtained by serial dilution and plate-counting. Briefly, this standard laboratory procedure [22] involved diluting the sample progressively over 9 orders of magnitude, then taking a 50 μl aliquot from each of them and spreading them over a petridish with Tryptic Soy Agar. The petridishes were incubated for ~48 hours, and the number of colonies was counted. Based on the colony counts from plates with 20–200 colonies, the number of colony forming units of bacteria present in the original 10 ml sample was estimated. The number of bacterial colony forming units present per unit surface area in our biofilm is obtained by dividing this number by the surface area of our coupon (biofilm).

In order to obtain the number of bacteria surviving our ultra-rapid combustion procedure, we collected the entire carbonaceous residue that we obtained at the end of the process, and dispersed it in 5 ml of PBS. We also scraped off the surface of interest, rinsed both the surface and the razor, and collected all the material together. As earlier, the volume was made up to 10 ml, the bacteria were dispersed using vortexing and sonication, and their number was estimated using serial dilution and plating. In many of the experiments (especially for the second and third sets of data that we obtained), we eschewed serial dilution during this part and obtained plate counts directly from the 10 ml of sample. Again, based on the number of colonies observed, an estimate of the number of viable bacteria present per unit area of the coupon (biofilm) surface was obtained.

RESULTS AND DISCUSSION

Estimates of the Numbers of Live Bacteria in the Biofilms Grown

The substrates, with and without biofilm, are shown in Figure 2. Also shown are scanning electron microscope (SEM) images of the biofilms. The number of viable bacteria per unit (nominal) area of these biofilms was obtained using the methods described in the previous section. These numbers are summarized in Table 2.

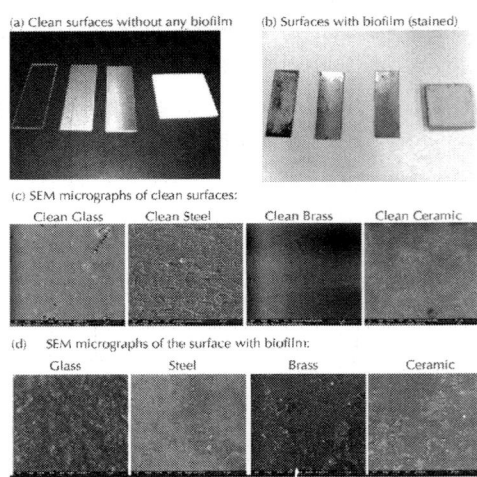

Figure 2: Images of substrates with and without biofilm and their corresponding SEM micrographs: (a) Photographs of the clean coupons before any bio-

film was grown on it, (b) Photograph showing the coupons after the growth of biofilm on the surface with Gram staining and (c) SEM micrographs the clean coupons before the growth of biofilms (Left to Right: Glass, Steel, Brass and Ceramic) and (d) SEM micrographs of the coupons with biofilm (Left to Right: Glass, Steel, Brass and Ceramic).

Table 2: Average, and standard deviation (n = 3) of the numbers of bacteria in the biofilms grown on various substrates

Substrate	Bacterial load in biofilm (mean ± standard deviation)
Glass	$1.86 \ (\pm 0.47) \times 10^7$
Ceramic	$1.26 \ (\pm 0.46) \times 10^8$
Brass	$5.43 \ (\pm 0.23) \times 10^7$
Steel	$1.20 \ (\pm 0.34) \times 10^8$

Lee et al.

Lee et al. BMC Biotechnology 2013 13:30   doi:10.1186/1472-6750-13-30

As seen, the numbers that we find are in the order of 10^7 - 10^8 CFU/cm^2. The SEM images also show approximately 25 to 35 bacteria in rectangular regions whose lengths and widths are less than 20 microns each, yielding estimates of $\sim 10^7$ - 10^8 bacteria/cm^2. These numbers also happen to be consistent with values reported elsewhere [2,11] for P.aeruginosa biofilms grown under moderate shear. Thus, our results also serve to verify that our method for growing biofilms is effective, and that our method for estimating bacterial surface densities is an acceptable one.

The Rapid Combustion Process

Pictures of the samples at various stages of the process are shown in Figure 3. (A video of the rapid combustion is also provided in the supplementary material). The first picture [Figure 3a] shows a coupon being ignited (after being spray-coated with the blend of fluoropolymer and aluminum nanoparticles dispersed in acetone, and allowing the acetone to subsequently evaporate). Subsequent images [Figure 3b and c] are taken 30 ms apart and as seen, the whole 3″ × 1″ top surface of the coupon burnt in < 100 ms. As recorded using a high speed camera, the flame propagated from the bottom-right to the top left corner, covering the diagonal length of the coupon (a distance of ~ 8 cm) in about 60 ms, thus yielding a linear propagation rate of about 1.3 m/s. As can been seen in Figure 4a, the combustion process left behind a carbonaceous residue, which was easily washed off from the surface to obtain the desired biofilm-free substrate (as shown in Figure 4b). Using SEM, we also examined the microstructure of the materials after it had been subjected to combustion (as shown in Figure 4c). On comparing the micrographs shown in Figure 4 to those in Figure 2, we were unable to observe any major damage or change to any of the materials.

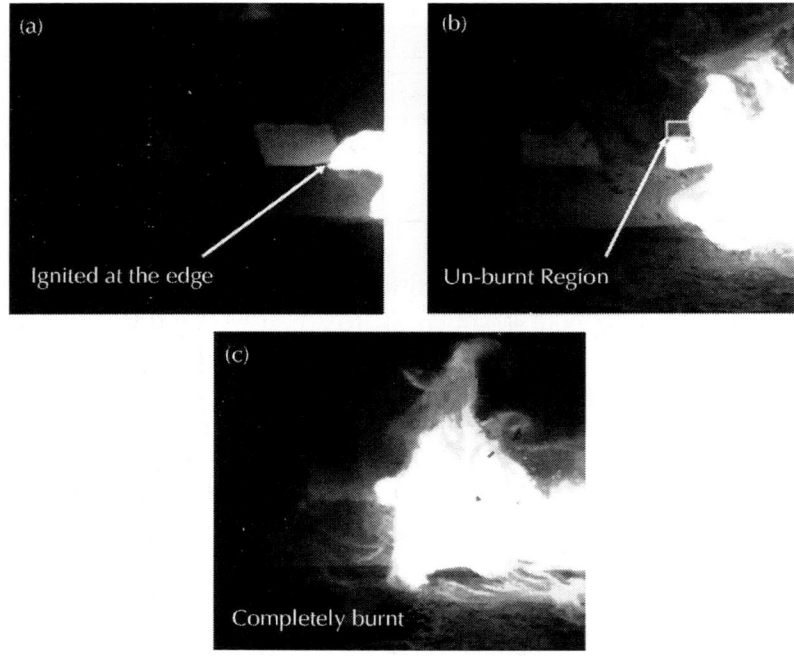

Figure 3: Substrate subjected to the novel process of ultra-rapid biofilm removal. The time arrival record of the flame front at different positions was determined using a high speed camera over a distance of 6 cm. (a) Initiation of the ignition at the edge of the coupon (time t = 0 ms) (b) The coupon with a small un-burnt region (time t = 30 ms) and (c) Surface of the coupon completely burnt (time t = 60 ms).

Figure 4: Images of substrates after burn and after wash with corresponding SEM micrographs: (a) Photograph of the coupons after the combustion process, showing the carbonaceous residue (b) Photograph showing the coupons after removal of the residue by washing (c) SEM Micrographs of the material surfaces after ultra-rapid biofilm combustion and (d) SEM micrographs of the coupons after removal of the carbonaceous residue by washing (Left to Right: Glass, Steel, Brass and Ceramic).

Estimates of the Efficacy of the Process

The number of viable bacteria present on the surface of the substrates (coupons) after the combustion (including those on the carbonaceous material, if any) is estimated using the method described earlier. This number also includes any bacteria present on the carbonaceous material (if any). These numbers (divided by the nominal area of the coupon surface to obtain the numbers per unit area), are plotted in Figure 5 along with the numbers present per unit area in the biofilm prior to the rapid combustion process.

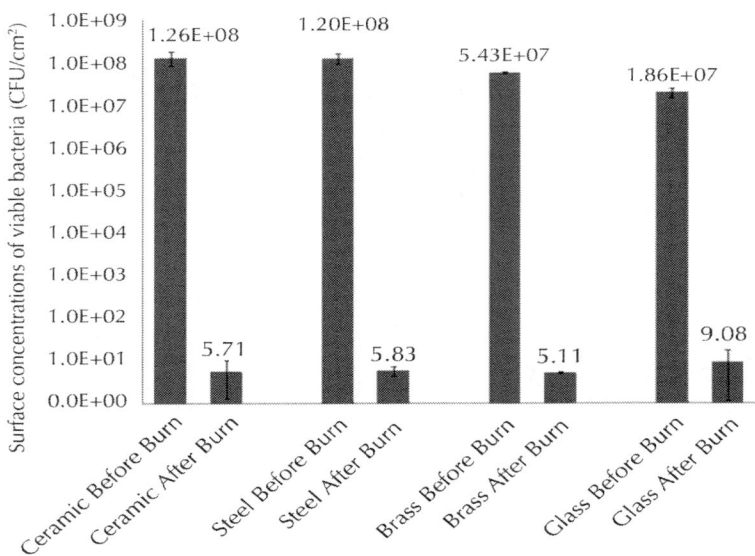

Figure 5: Efficacy of our ultra-rapid process for the removal of biofilms: The graph shows the average (n = 4) and standard deviations for the surface concentration of viable bacteria (CFU/cm2) on different materials before and after the combustion process.

As seen in Figure 5, the method developed in the present work is able to reduce the numbers of bacteria on the surfaces to ~10 CFU/cm². This represents a 5 to 6 $\log_{10}$ decrease from its original concentration. It may also be noted that the rapid combustion process, and the subsequent collection of carbonaceous material was performed in an environment that was not sterile, and hence random bacteria from the environment were likely to have been introduced into our "after" sample, leading to the estimate being higher than the true value of viable bacteria remaining.

In Table 1, we compared the efficiency of the rapid combustion process developed in this work to a few other methods reported in literature for eliminating bacteria / biofilms on solid substrates on two criteria: the degree to which the process is able to reduce the infestation of bacteria (as determined by the log-reduction in the number of viable bacteria), and the time taken to carry out the process. We compare our process of biofilm removal only to others that are used to treat process equipment or surfaces "offline", and we do not foresee our combustion process being used to eliminate biofilms while a chemical or biological process is still ongoing. (The latter category includes biofilms on the surfaces of implanted medical devices such as orthopedic joints, pacemakers etc.)

As seen, the performance of our process is better that of the other methods listed on one or both counts. Firstly, its efficacy in eliminating viable bacteria harbored in biofilms is really high (we obtain a > 5 log reduction, as compared to 2–3 logs for most other processes), and secondly, its turn-around time is really short, with the core combustion process taking less than 1 second and the prior preparation steps (spraying) taking about 1 minute. Taken together, these two features make it an extremely promising commercial technology, especially for applications involving "offline" removal of biofilms from metallic and ceramic surfaces of process equipment, where both efficacy of removal and turn-around time are the key considerations. The other advantage of our method is that it generates very little solid or liquid wastes (the disposal of which typically requires additional resources to be expended).

While the experiments reported here demonstrate the potential of this approach, much work remains to be done before we can claim that it is ready for use in the real world. For instance, though we suspect that the carbonaceous residue that we obtained post-combustion may contain traces of the substrate material (metal or ceramic), quantification of these trace elements is beyond the scope of the current work. Depending on the amount of material lost, our process may be deemed acceptable for certain applications, and unacceptable for others. Also, we are not sure if any aerosols containing live bacteria are released during the process. If found to be so, additional precautions may be needed prior to using our method in real world situations. In addition, we will have to demonstrate that the process can successfully remove the older, more complex, mixed culture biofilms that are usually seen in process equipment, and investigate how the presence of chemical (as opposed to biological) fouling products (such as rust) affect the performance of this process. Additional issues that we will have to consider include formulating a blend that remains safe to handle even when used in large quantities and under challenging conditions, and formulating slower burning and/or less high-temperature blends for polymeric/plastic surfaces.

CONCLUSIONS

The method of biofilm removal via the combustion of a nanoenergetic coating reported in this work is presently in an early stage of development. While much work remains before the method can be used in "real-world" settings, the present work nevertheless provides the scientific community with essential information regarding this novel method.

AUTHORS' CONTRIBUTIONS

The nanoparticle-blend was formulated and prepared by RT and BMS, with guidance from KG and SG. The biofilm growth, ablation and evaluation were done by BDL, SP and BMS, with the guidance of SS. SS directed the overall study. All authors approve the final manuscript.

ACKNOWLEDGEMENTS

BDL was supported in part by the Cheongbung Research Foundation, South Korea. BMS and SP were supported from Startup funds issued to SS by the Dept. of Biological Engineering, University of Missouri.

REFERENCES

1.   Donlan RM, Costerton JW: Biofilms: survival mechanisms of clinically relevant microorganisms.Clin Microbiol Rev 2002, 15:167-193.

2.   Goeres DM, Loetterle LR, Hamilton MA, Murga R, Kirby DW, Donlan RM: Statistical assessment of a laboratory method for growing biofilms.Microbiology 2005, 151:757-762.

3.   Ghigo J-M: Natural conjugative plasmids induce bacterial biofilm development.Nature 2001, 412:442-445.

4.   Holm E, Schultz M, Haslbeck E, Talbott W, Field A: Evaluation of hydrodynamic drag on experimental fouling-release surfaces, using rotating disks.Biofouling 2004, 20:219-226.

5. Mahfouz AB, El-Halwagi MM, Batchelor B, Atilhan S, Linke P, Abdel-Wahab A: Process integration and systems analysis for seawater cooling in industrial facilities. InDesign for Energy and the Environment. Edited by El-Halwagi MM, Linninger AA. Boca Raton, FL: CRC Press; 2010:223-232.

6. Melo L, Bott T: Biofouling in water systems.Experimental Thermal and Fluid Science 1997, 14:375-381.

7. Ludensky M: Control and monitoring of biofilms in industrial applications.Int Biodeter Biodegr 2003, 51:255-263.

8. Maxwell S, Devine C, Rooney F: Monitoring and Control of Bacterial Biofilms in Oilfield Water Handling Systems. In Book Monitoring and Control of Bacterial Biofilms in Oilfield Water Handling Systems. New Orleans, LA: NACE International; 2004.

9. Elsmore R, Gilbert P, Ludensky M, Coulbrum J: Biofilms: the Good, the Bad and the Ugly. Edited by Wimpenny J, Gilbert P, Walker J, Brading M, Bayston R. Cardiff: Bioline; 1999:81-89.

10. Muller-Steinhagen H: Heat Exchanger Fouling: Mitigation and Cleaning Techniques. UK: Inst of Chemical Engineers; 2000.

11. Gibson H, Taylor JH, Hall KE, Holah JT: Effectiveness of cleaning techniques used in the food industry in terms of the removal of bacterial biofilms.J Appl Microbiol 1999, 87:41-48.

12. Chen X, Stewart PS: Biofilm removal caused by chemical treatments.Water Res 2000, 34:4229-4233.

13. Orgaz B, Neufeld RJ, SanJose C: Single-step biofilm removal with delayed release encapsulated Pronase mixed with soluble enzymes.Enzyme Microb Technol 2007, 40:1045-1051.

14. Oulahal N, Martial-Gros A, Bonneau M, Blum LJ: Combined effect of chelating agents and ultrasound on biofilm removal from stainless steel surfaces. Application to "Escherichia coli milk" and "Staphylococcus aureus milk" biofilms.Biofilms 2004, 1:65-73.

15. Karau MJ, Piper KE, Steckelberg JM, Kavros SJ, Patel R: In vitro activity of the qoustic wound therapy system against planktonic and biofilm bacteria.Adv Skin Wound Care 2010, 23:316-320.

16. Belnkinsopp SA, Khoury AE, Costerton JW: Electrical enhancement of biocide efficacy against pseudomonas aeruginosa biofilms. Appl Environ Microbiol 1992, 58:3770-3773.

17.   Kang M-Y, Jeong H-W, Kim J, Lee J-W, Jang J: Removal of biofilms using carbon dioxide aerosols.Journal of Aerosol Science 2010, 41:1044-1051.

18.   Beckstead MW: A Summary of Aluminum Combustion.RTO/VKI Special Course on "Internal Aerodynamics in Solid Rocket Propulsion"; Rhode-Saint-Genese, Belgium, Volume RTO-EN-023 2002, 5.1-5.46.

19.   Holley CEJ, Huber EJJ: The heats of combustion of magnesium and aluminum.J Am Chem Soc 1951, 73:5577-5579.

20.   Costerton JW, Stewart PS, Greenberg EP: Bacterial biofilms: a common cause of persistent infections.Science 1999, 284:1318-1322.

21.   Shende R, Subramanian S, Hasan S, Apperson S, Thiruvengadathan R, Gangopadhyay K, Gangopadhyay S, Redner P, Kapoor D, Nicolich S, Balas W: Nanoenergetic composites of CuO nanorods, nanowires, and Al-nanoparticles.Propellants, Explosives, Pyrotechnics 2008, 33:122-130.

22.   Maturin L, Peeler JT: Aerobic Plate Count. In Bacteriological Analytical Manual. 8th edition. Edited by Jackson GJ, Merker RI, Bandler R. Gaithersburg, MD: AOAC International; 1998.

Chapter 2

Thermal-Hydraulic System Codes in Nulcear Reactor Safety and Qualification Procedures

Alessandro Petruzzi and Francesco D'Auria

DIMNP, University of Pisa, Via Diotisalvi 2, Pisa 56100, Italy

ABSTRACT

In the last four decades, large efforts have been undertaken to provide reliable thermal-hydraulic system codes for the analyses of transients and accidents in nuclear power plants. Whereas the first system codes, developed at the beginning of the 1970s, utilized the homogenous equilibrium model with three balance equations to describe the two-phase flow, nowadays the more advanced system codes are based on the so-called "two-fluid model" with separation of the water and vapor phases, resulting in systems with at least six balance equations. The wide experimental campaign, constituted by the integral and separate effect tests, conducted under the umbrella of the OECD/CSNI was at the basis of the development and validation of the thermal-hydraulic system codes by which they have reached the present high degree of maturity. However, notwithstanding the huge amounts of financial and human resources invested, the results predicted by the code are still affected by errors whose origins can be attributed to several reasons as model deficiencies, approximations in the numerical solution, nodalization effects, and imperfect knowledge of boundary and initial conditions. In this context, the existence of qualified procedures for a consistent application of qualified thermal-hydraulic system code is necessary and implies the drawing up of specific criteria through which the code-user, the nodalization, and finally the transient results are qualified.

INTRODUCTION

Evaluation of nuclear power plants (NPPs) performances during accident conditions has been the main issue of the research in nuclear fields during the last 40 years. Therefore, several complex system thermal-hydraulic codes have been developed for simulating the transient behavior of water-cooled reactors. In the early stage of the development, the codes were primarily applied for the design of the engineered safety systems. In 1978, the "appendix K requirements" [1] were issued, defining conservative model assumptions as well as conservative initial and boundary conditions to warrant conservative code results for critical safety parameters. On the other hand, the development and elaboration of accident management procedures, the application of probabilistic safety analyses (PSA) and the operator training asked for so-called "best-estimate (BE) analysis," that means an accident simulation as realistic as possible. The main objective of best-estimate system codes was to replace the "evaluation models," which used many conservative assumptions, by the best-estimate approach for more realistic predictions of pressurized water reactor (PWR) or boiling water reactor (BWR) accidental transients that allow the reduction of safety margins. Best-estimate system codes are currently used for the following:

- safety analysis of accident scenarios;
- quantification of the conservative analyses margin;

- licensing purposes if the code is used together with a methodology to evaluate uncertainties;
- probabilistic safety analysis (PSA);
- development and verification of accident management procedures;
- reactors design;
- analysis of operational events;
- core management investigation.

Best-estimate thermal-hydraulic codes (e.g., RELAP, TRAC, CATHARE, ATHLET, ...) are, in general, based on equations for two-phase flow which are typically resolved in Eulerian coordinates. The two-phase flow field is described by mass, momentum, and energy conservation equations for the liquid and vapour phases separately and mass conservation equations for noncondensable gas present in the mixture. The models are suitable for 1D system simulation even if for some NPP component (e.g., the vessel), some code has the capability to solve 3D system equations. Time discretization could be fully, semi or nearly implicit. Depending on the number of balance equations, different sets of constitutive equations are required to close the equation system. In comparison with the homogeneous equilibrium model (HEM), which requires only two constitutive equations, namely, the friction loss and the heat transfer relations at the wall, at least seven constitutive equations are required for the two fluid models with six balance equations describing the mass, energy, and momentum transfers at the interface and the energy and momentum transfers of the water- and steam-phase at the wall. The constitutive equations have to describe the physical phenomena in a wide span of scale, ranging from down-scaled integral system experiments up to full size reactor geometry. This is one of the most challenging goals in code development and code validation. To develop and validate the scaling laws for individual phenomena, separate effect tests in different scale are necessary. In Figure 1, the code development activities carried out in more than three decades are shown.

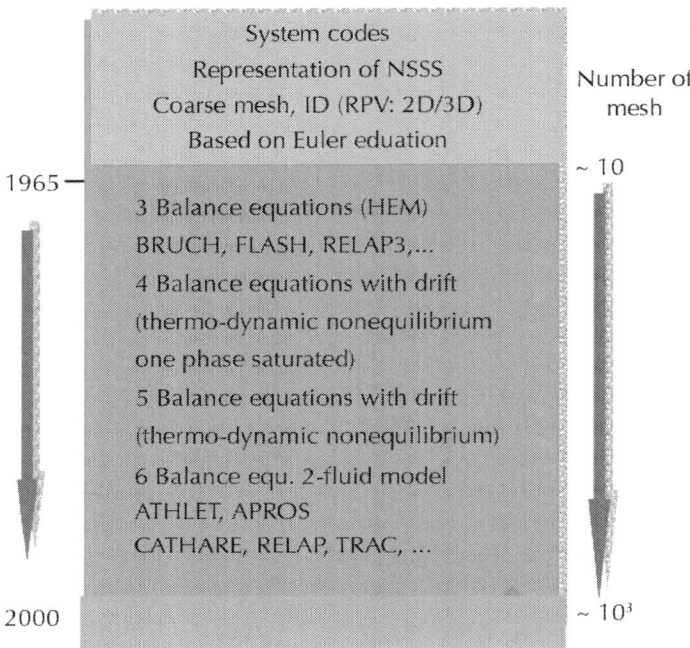

Figure 1: Code development activities in more than three decades.

Due to the numerical approximations and the empirical nature of the included models in the thermal-hydraulic system codes, extensive activities related to validation of the codes have been pursued during the years. The validation has been performed using experimental data from specially designed scaled-down test facilities. In addition, transient data from real NPPs were also considered due to the full scale and true geometry although those data concern only conditions under fairly mild transients (operational transients and start-up and commissioning tests). These activities have been planned and carried out in national and international contexts in four levels, mainly in the independent assessment area, involving the use of the following:

"fundamental" experiments [2];

- separate effects test facilities (SETF) [3];
- integral test facilities (ITF), including most of the international standard problems (ISP) [4];

- real plant data.

However, notwithstanding the huge amounts of financial and human resources invested, the results predicted by the code are still affected by errors whose origins can be attributed to several reasons as model deficiencies, approximations in the numerical solution, nodalization effects, and imperfect knowledge of boundary and initial conditions. In this context, the existence of qualified procedures for a consistent application of qualified thermal-hydraulic system code is necessary and implies the drawing up of specific criteria through which the code-user, the nodalization, and finally the transient results are qualified.

The current situation related to the development, validation, and use of system codes can be summarized as follows.

- A state-of-the-art report in modeling LOCA (loss-of-coolant accident) and non-LOCA transients and the compendium on ECCS (emergency core cooling systems). Researches have been published in 1989 [5, 6], by Organization for Cooperation and Development/Committee on the Safety of Nuclear Installations (OECD/CSNI) and US NRC. These reports broadly cover topics like plant features relevant to thermal-hydraulics, transient description, phenomena identification, code modeling capabilities and needs for experimental data and present situation in the experimental area.

- The CSAU (Code Scaling, Applicability and Uncertainty), published in 1990, for example [7], constituted a pioneering effort made by NRC in the area of code uncertainty prediction.

- Code validation criteria and detailed qualification programs exist, although not fully optimized or internationally agreed on. In particular, the following hold.

a. The integral test facility CSNI code validation matrix (ITF-CCVM) report was initially published in 1987 and extensively updated in 1996, [4]. Tests for code validation were selected based on quality of the data, variety of scaling and geometry, and appropriateness of the range of covered conditions. The decision was taken around 1984 to bias the validation matrix toward integral tests so that code models were exercised and interacted in situations as similar as possible to those of interest to PWR and BWR. This was done because of the assumption that sufficient comparison with separate effects test data would be performed and documented by code developers.

b.   As the last expectation has proved unrealistic, a group of scientists was formed toward the end of the 80s to set up the separate effect test facility CSNI code validation matrix, SETF-CCVM, that was issued in 1994 [3]. The development of the SETF-CCVM required an extension of the methodology employed for the ITF-CCVM [4], both in the scope and the definition of the thermal-hydraulic phenomena and in the categorization and description of facilities. A significant result of the activity was the selection of sixty-seven phenomena assumed to cover all the thermal-hydraulic situations of interest expected in PWR and BWR transients. The needed effort suitable for a comprehensive code validation was quantified: more than one thousand experiments should be part of a thermal-hydraulic system code validation program. The impact of those findings in planning new researches was also evaluated [8].

- The codes have reached an acceptable degree of maturity although the reliable application is still limited to the validation domain.

- The use of qualified codes is more and more requested for assessing the safety of existing reactors, especially in the former Soviet Union and in the Eastern countries, and for designing advanced reactors.

- The codes availability is increasingly growing especially in the countries belonging to the former Soviet Union, the Eastern countries, Korea, China, and so forth.

- Special topics, like user [9] and computer-compiler effects upon code calculation results, nodalization qualification [10], accuracy quantification [11], relevance of international standard problems and lesson learned, use of best estimate codes in the licensing, have been widely discussed and main achievements are available to the international community.

- A special attention from the scientific community has always been given to the quantification of code uncertainty in predicting plant transients. Methodologies to evaluate the "uncertainty" have been proposed [12, 13] and tested in several international activities, like UMS (uncertainty method study, [14]) and BEMUSE (best-estimate methods–uncertainty and sensitivity evaluation, [15, 16]) that allowed the comparison of uncertainty results obtained from different methodologies.

This paper reviews the main features and limitations of the thermal-hydraulic system codes and the procedures adopted for the qualification of computational tools, that is, not only the codes, through the ITF and SETF validation matrixes, but also the nodalization used to simulate the transient scenario in the NPP. Finally, taking into account the multidisciplinary nature of reactor transients and accidents (which include thermal-hydraulics, neutronics, structural, and radiological aspects), the needs, the status of development, and the benefits of code coupling are pointed out.

MAIN FEATURES AND LIMITATIONS OF THERMAL-HYDRAULIC SYSTEM CODES

The system thermal-hydraulic codes are based upon the solution of six balance equations for liquid and steam that are supplemented by a suitable set of constitutive equations. The balance equations are coupled with conduction heat transfer equations and with neutron kinetics equations (typically point kinetics). The two-phase flow field is organized in a number of lumped volumes connected with junctions. Thermal-hydraulic components such as valves, pumps, separators, annulus, accumulators, and so forth, can be defined in order to represent the overall system configuration. In the following sections, main problematic aspects, from the point of view of the user, of a thermal-hydraulic system code are highlighted.

System Nodalization

All major existing light water reactor (LWR) safety thermal-hydraulics system codes follow the concept of a "free nodalization," that is, the code user has to build up a detailed noding diagram which maps the whole system to be calculated into the frame of a one-dimensional thermal-hydraulic network. To do this, the codes offer a number of basic elements like single volumes, pipes, branches, junctions, heat structures, and so forth. This approach provides not only a large flexibility with respect to different reactor designs, but also allows predicting separate effect and integral test facilities which might deviate considerably from the full-size reactor.

As a consequence of this rather "open strategy," a large responsibility is passed to the user of the code in order to develop an adequate nodalization scheme which makes best use of the various modules and the prediction capabilities of the specific code. Due to the existing code limitations and to economic constraints, the development of such a nodalization represents always a compromise between the desired degree of resolution and an acceptable computational effort. It is not possible here to cover all the aspects of the development of an adequate nodalization diagram, however, two crucial problems will be briefly mentioned which illustrate the basic problem.

Spatial Convergence

As has been quite often misunderstood, a continuous refinement of the spatial resolution (e.g., a reduction of the cell sizes) does not automatically improve the accuracy of the prediction. There are two major reasons for this behavior:

- the large number of empirical constitutive relations used in the codes has been developed on the basis of a fixed (in general coarse) nodalization;
- the numerical schemes used in the codes generally include a sufficient amount of artificial viscosity which is needed in order to provide stable numerical results. A reduction of the cell sizes below a certain threshold value might result in severe nonphysical instabilities.

From those considerations, it can be concluded that no a priori optimal approach for the nodalization scheme exists.

Mapping of Multidimensional Effects

Multidimensional effects, especially with respect to flow splitting and flow merging processes (e.g., the connection of the main coolant pipe to the pressure vessel), exist also in relatively small scale integral test facilities. The problem might become even more complicated due to the presence of additional bypass flows and a large redistribution of flow during the transient. It is left to the code user to determine how to map these flow conditions within the frame of a one-dimensional code, using the existing elements like branch components, multiple junction connections, or cross-flow junctions. These two examples show how the limitations in the physical modeling and the numerical method in the codes have to be compensated by an "engineering judgment" of the code user which, at best, is based on results of detailed sensitivity of assessment studies. However, in many cases, due to lack of time or lack of appropriate experimental data, the user is forced to make ad hoc decisions.

Code Options: Physical Model Parameters

Even though the number of user options has been largely reduced in the advanced codes, various possibilities exist about how the code can physically model specific phenomena. Some examples are as follows.

- Choice between engineering type models for choking or use of code implicit calculation of critical two-phase flow conditions.
- Flow multipliers for subcooled or saturated choked flow.
- The efficiency of separators.
- Two-phase flow characteristics of main coolant pumps.
- Pressure loss coefficient for pipes, pipe connections, valves, branches, and so forth.

Since in many cases direct measured data are not available or, at least, not complete, the user is left to his engineering judgment to specify those parameters.

Input Parameter Related to Specific System Characteristics

The assessment of LWR safety codes is mainly performed on the basis of experimental data coming from scaled integral or separate effect test facilities. Typically in these scaled-down facilities, specific effects, which might be small or even negligible for the full-size reactor case, can become as important as the major phenomena to be investigated.

Examples are the release of the heat from the structures to the coolant, heat losses to the environment, or small bypass flows. Often, the quality of the prediction depends largely on the correct description of those effects which needs a very detailed representation of the structural materials and a good approximation of the local distribution of the heat losses. However, many times the importance of those effects is largely underestimated, and consequently, wrong conclusions are drawn from results based on incomplete representation of a small-scale test facility.

Input Parameters Needed for Specific System Components

The general thermal-hydraulic system behavior is described in the codes by the major code modules based on a one-dimensional formulation of the mass, momentum, and energy equations for the separated phases. However, for a number of system components, this approach is not adequate and consequently additional, mainly empirical models have to be introduced, for example, for pumps, valves, separators, and so forth. In general, these models require a large amount of additional code input data, which are often not known since they are largely scaling dependent.

A typical example is the input data needed for the homologous curves which describe the pump behavior under single and two-phase flow conditions which in general are known only for a few small-scale pumps. In all these cases, the code user has to extrapolate from existing data obtained for different designs and scaling factors which introduces a further uncertainty to the prediction.

Specification of Initial and Boundary Conditions

Most of the existing codes do not provide a steady-state option. In these cases pseudo-steady-state runs have to be performed using more or less artificial control systems in order to drive the code towards the specified initial conditions. The specification of stable initial and boundary conditions and the setting of related controllers require great care and detailed checking. If this is not done correctly, a large risk, that even small imbalances in the initial data will overwrite the following transient, exists especially for slow transients and small break LOCA calculations.

Specification of State and Transport Property Data

The calculation of state and transport properties is usually done implicitly by the code. However, in some cases, for example, in RELAP5, the code user can define the range of reference points for property tables, and therefore, can influence the accuracy of the prediction. This might be of importance especially in more "difficult regions," for example, close to the critical point or at conditions near atmospheric pressure. Another example is constituted by the fuel materials property data: the specification of fuel rod gap conductance (and thickness) is an important parameter, affecting core dryout and rewet occurrences that must be selected by the user.

Selection of Parameters Determining Time Step Sizes

All the existing codes are using automatic procedures for the selection of time step sizes in order to provide convergence and accuracy of the prediction. Experience shows, however, that these procedures do not always guarantee stable numerical results, and therefore, the user might often force the code to take very small time steps in order to pass through trouble spots. In some cases, if this action is not taken, very large numerical errors can be introduced in the evolution of any transient scenario and are not always checked by the code user.

Code Input Errors

In order to prepare a complete input data deck for a large system, the code user has to provide a huge number of parameters (approx., 15 to 20 thousand values for an NPP nodalization) which he has to type one by one. Even if all the codes provided consistency checks, the probability for code input errors is relatively high and can be reduced only by extreme care following clear quality assurance guidelines.

QUALIFICATION OF COMPUTATIONAL TOOLS

A key feature of the activities performed in nuclear reactor safety technology is constituted by the necessity to demonstrate the qualification level of each tool adopted within an assigned process and of each step of the concerned process. Computational tools include (numerical) codes, nodalizations, and procedures. Furthermore, the users of those computational tools are part of the process and need suitable demonstration of qualification.

A consistent application (development, qualification, and application) of a thermal-hydraulic system code is depicted in Figure 2. The code development and improvement process, block 1 in Figure 2, is conducted by "code developers" who make extensive use of assessment (block 4), typically performed by independent users of the code (i.e., group pf experts independent from those who developed the code). The consistent code assessment process implies the availability of experimental data and of robust procedures for the use of the codes, blocks 2 and 3, respectively. Once the process identified by blocks 1 and 4 is completed, a qualified code is available to the technical community, ready to be used for NPP applications (block 5). The NPP applications still require "consistent" procedures (block 3) for a qualified use of the code. The results from the calculations are, whatever the qualification level achieved by the code is, affected by errors that must be quantified through appropriate uncertainty evaluation methodology (block 6).

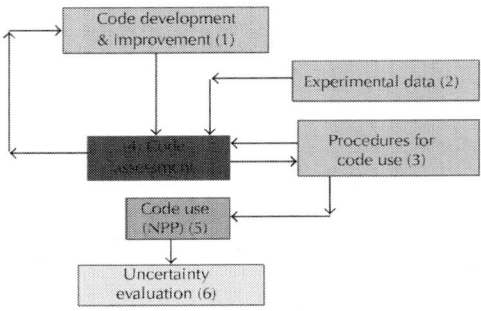

Figure 2: A consistent application (development, qualification and application) of a thermalhydraulic system code.

Code Qualification

The code constitutes the main tool for investigating the NPP behavior or for evaluating the efficacy of systems or special procedures during accident transient scenarios. The following constitutes the main requisites for a qualified use of the code [11].

- Capability of the code to reproduce the relevant phenomena occurring for the selected spectrum of accidents.
- Capability to reproduce the peculiarities of the reference plant/ facility.
- Capability to produce suitable results for a comparison with the acceptable criteria.
- Availability of qualified users.

Essentially the code must be able to reproduce two fundamental aspects [17].

a. The NPP and the accident conditions: all the relevant zones, systems, procedure, and related actuation logic is to be included in the calculation. This item also includes any external event, boundary and initial condition necessary to identify the plant but also the selected accident.

b. The phenomena occurring (expected) during the accident.

In order to ensure those capabilities, the code qualification process is needed and the following two phases can be identified.

- Development phase: several models are created, developed, and improved by the code development team; many checks are necessary to qualify each model and the global architecture of the code.
- Independent assessment phase: the code is ready to be used but qualified calculations performed by organizations independent from the code-development team are needed to check independently the declared capabilities of the code.

It is relevant to note that in the development phase the code models can be changed and the code is not available to the final user. In the independent assessment phase, the final version of the code is distributed and the user is generally forbidden to change any element of the code models apart from the normal available options as described in the user manual.

The activities performed during the development phase are (Figure 3) as follows.

a. Verification: it consists in the review of the source coding relative to its description in the documentation. In other words, code verification involves activities that are related to software quality assurance (SQA) practices and to activities directed toward finding and removing deficiencies in models and in numerical algorithms used to solve partial differential equations. SQA procedures are needed during software development and modification, as well as during production computing. SQA procedures are well developed in general, but areas of improvement are needed with regard to software operating on massively parallel computer systems. During the verification step, the correct working of models, interfaces, and numerics is checked to ensure that the code, in all its components, is free of errors and produces results.

b. Validation (or assessment): it consists in evaluating the accuracy of the values predicted by the code-nodalization against relevant experimental data for important phenomena expected to occur. In other words, code validation emphasizes the quantitative assessment of computational model accuracy by comparison with high-quality validation experiments, that is, experiments that are well characterized in terms of measurement and documentation of all the input quantities needed for the computational model, as well as carefully estimated and documented experimental measurement uncertainty. The validation process ensures the consistency of the results produced by the code; that is, it proves that the code, as a whole system, is capable to produce meaningful results: not only the code-system works, but it also works in the right direction.

The independent code-assessment is carried out by independent users of the code and has the aim to quantify the code accuracy, which is the discrepancy between transient calculations and experiments performed in ITF. The independent assessment of the code involves different aspects, like (Figure 3)

•   qualification of the nodalization;
•   qualification of the user;
•   definitions of procedures for the use of the code;

- evaluation of the accuracy from a qualitative and quantitative point of view.

The above items are connected with the application of the code to experimental tests performed in ITF. The procedure for the qualification of the nodalization is described with more details in the Section 3.4 together with acceptability criteria.

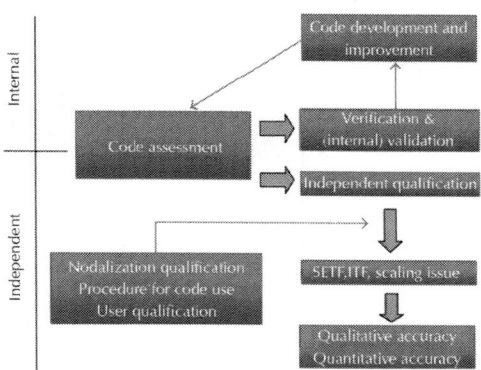

Figure 3: Internal and external (independent) code assessment.

Besides the demonstration of the code capability in reproducing an experiment performed in a test facility, the code must be checked also in performing NPP calculation. This constitutes the final step of the independent code assessment (Figure 4): the demonstration of the code capability at a different scale, that is, the full scale of the NPP. A nodalization of an NPP is prepared and qualified. The check consists in a "similarity analysis" generally involving a Kv-scaled calculation (see Section 3.3). In this kind of calculation, the initial and boundary conditions of an experiment performed in an ITF are properly scaled and implemented in the NPP nodalization. The results of the NPP-scaled nodalization must reproduce the relevant phenomena occurring in the experiment. Alternative ways to prove the code capability at the NPP scale are constituted by the comparison with other qualified NPP code results or, if available, with data obtained in NPP operational transients. As the procedure followed for this part of the code assessment is the same adopted for the qualification process of the nodalization, more details are given in Section 3.4.

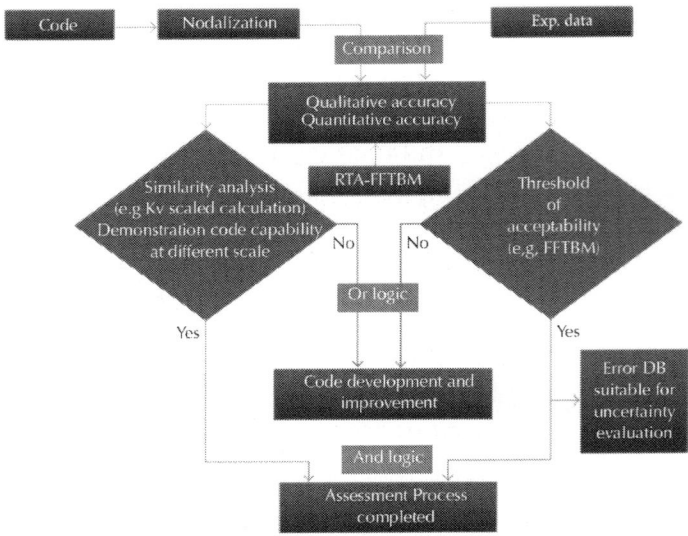

Figure 4: Code independent assessment.

The contemporaneous acceptability of the accuracy (step of the process connected with experiments in ITF) and of the similarity analysis (step of the process connected with NPP) constitutes the positive demonstration of the code capability and the end of the code assessment. The calculated accuracy is possibly included in the data base suitable for uncertainty evaluation (block 6 in Figure 2, [12, 13]). If the accuracy is not in the range of acceptability or the code fails the similarity analysis, the code is considered not qualified and the code-development team will be informed in order to develop new code models or to improve the existing ones.

As consequence, new revision or new version of the code can be produced during the development phase: a new revision contains a new physical modeling whereas a new version may contain new numerical methods, new modules, new submodules, new preprocessing or post-processing or a new code architecture. The steps typically performed during the qualification process of a new revision or of a new version of the code are depicted in Figure 5. The needed reference data are derived by the following sources.

- Analytical experiments, with separate effect tests and component tests, are used for the development and the validation of closure laws.

- System tests or integral tests used to validate the general consistency of the revision. Successive revisions of constitutive laws are implemented in successive versions of the code and assessed.

Constitutive relationships are developed and assessed following a general methodology hereafter summarized.

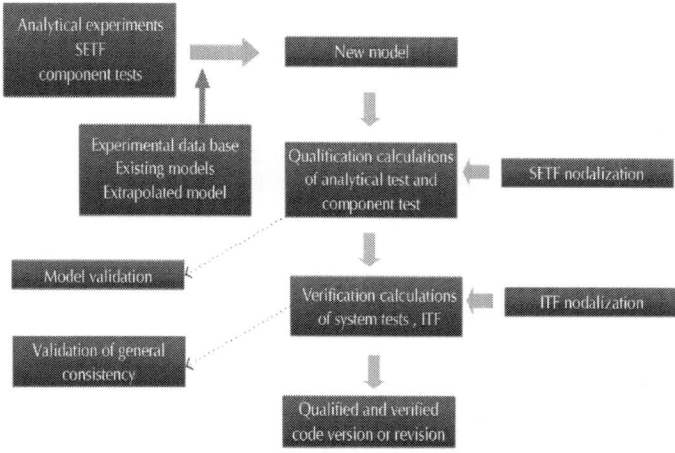

Figure 5: Qualification process of a new revision or a new version of the code.

Step A

Analytical experiments, including separate effect tests and component tests, are performed and analyzed. Separate effect tests investigate a physical process such as the interfacial friction, the wall heat transfer. Component tests investigate physical processes which are specific to a reactor component, such as the phase separation in a Tee junction.

Step B

Development of a complete revision of constitutive laws from a large analytical experimental data base. Successive revisions are implemented in successive code versions.

Step C

Qualification calculations of the analytical tests are used in order to validate each closure relationship.

Step D

Verification calculations of system tests or integral tests are used in order to validate the general consistency of the revision.

Step E

Delivery of the code version and revision is fully assessed (qualified and verified) and documented (description documents and assessment reports).

A new revision of constitutive laws is developed using some general principles.

- Data are first compared with existing models; if necessary, original models are developed.
- When and where data are missing, simple extrapolations of existing qualified models are used. No mechanistic model is developed without the experimental evidence of its relevance.
- In a prequalification phase, some tests of each experiment of the qualification matrix are calculated.
- A systematic qualification of the frozen revision is then performed. All tests of the qualification matrix are calculated and qualification reports are written.

Some other additional remarks about the qualification process of the code are as follows.

- The qualification program has to cover the whole range of accidental transients in LWR. As examples, the following accidents have to be considered for a PWR: large break loss of coolant accidents (LBLOCA); small break loss of coolant accidents (SBLOCA); steam generator tube ruptures (SGTR); loss of feed water (LOFW); main stream line break (MSLB); loss of residual heat removal (RHR) system.

- The code has to be fully portable on all machines, so that a unique code version is released to all the users.
- No code options for physical models, or as few as possible, have to be proposed to the user.
- The users guidelines should be as precise as possible and take full benefit of the experience gained from the code-development team.

Validation Activities for Thermal-Hydraulic System Codes

The validation against experimental data is essential in the process of system codes development and improvement as it has been discussed in the previous section. The models implemented and used in a code are generally developed based on experimental tests performed in specific facilities. It is possible to distinguish among.

- Basic facilities: In these facilities the fundamental phenomena are reproduces; the results are used to improve the equations of the single model or to derive empirically the relation between the relevant parameters; this kind of facilities are designed with goal to reproduce the specific phenomenon to be investigate.
- Separate effect facilities: in these facilities some relevant zones of the NPP are reproduced by a suitable scaling law to investigate the local occurrence of a phenomenon; the results of the experiments performed in these facilities are used to create and to validate the (several) models to be included in a code.
- Integral tests facilities: these facilities are simulators of reference NPP. All the relevant parts and systems of an NPP are reproduced by a suitable scaling law. The whole plant is reproduced and the global plant response is obtained as results. The results are used to realize and improve the models and to check the code capabilities.

It will be noted that also the data from NPP can be used, if available. However, in an NPP the data obtained are the one recorded by the system of control of the plant while, typically, the facilities are equipped with a large number of sensors and many detailed data are generated making the instrumentation of the facilities more suitable for code validation.

Huge effort was done by the OECD/NEA/CSNI from 1991 to 1997 in the construction of the separate effects test facility code validation matrix (SETF-CCVM, published in 1994) for thermal-hydraulic system codes [3]. Integral test facility (ITF) matrices for validation of realistic thermal-hydraulic system computer codes were also established by CSNI focused mainly on PWRs, and BWRs. The ITF-CCVM [4] validation matrix was issued in 1987 and updated in 1996.

By the validation matrices, the best sets of openly available experimental data for code validation, assessment, and improvement were collected in a systematic way. Quantitative code assessment with respect to the quantification of uncertainties in the modeling of individual phenomena by the codes is also an outcome of the matrix development. In addition, the construction of such matrices is an attempt to record information of the experimental work which has been generated around the world over the last years in the LWR safety thermal-hydraulics field. 187 facilities covering 67 relevant phenomena for LOCA and non-LOCA transient applications of PWRs and BWRs within a large range of useful parameters were identified and about 2094 tests were included in the SETF-CCVM matrix. The majority of these phenomena are also relevant to advanced water-cooled reactors. The major elements of the SETF-CCVM have been already integrated into the validation matrices of the major best-estimate thermal-hydraulic system codes, for example, RELAP5, CATHARE, TRACE, and ATHLET.

A total number of 177 PWR and BWR integral tests have been selected as potential source for thermal-hydraulic code validation in the ITF-CCVM report. Counter-part tests, similar tests and OECD ISP tests were introduced in the report. Counter-part tests and similar tests in differently scaled facilities are considered highly important for code validation and therefore they were included in the tables of ITF selected experiments. Moreover, over the last twenty-nine years, CSNI has promoted 48 ISPs [18]. The main objectives of the ISPs are as follows: to contribute to better understanding of postulated events, to compare and evaluate the capability of codes (mainly best estimate codes), to suggest improvements to the code developers, to improve the ability of code users and to address the so-called scaling effect.

ISPs were performed in different fields as in-vessel thermal-hydraulic behavior, fuel behavior under accident conditions, fission product release and transport, core/concrete interactions, hydrogen distribution and mixing, containment thermal-hydraulic behavior. ISP experiments were carefully controlled, documented, and evaluated.

Addressing the Scaling Issue

The reason why this section has been included in the paper directly derives from the fact that the scaling analysis is the needed link between the experiments performed in ITF and SETF and their utilization in the code validation process. The flow diagram in Figure 6 emphasizes this relevant role of the scaling analysis (red boxes) in two different parts of the process describing a consistent application (development, qualification, and application) of a thermal-hydraulic system code: firstly during the code assessment process (as the code development and improvement is based on experimental data obtained in test facilities), secondly during the demonstration of the qualification of an NPP nodalization (which is a needed step to perform a reliable NPP calculation).

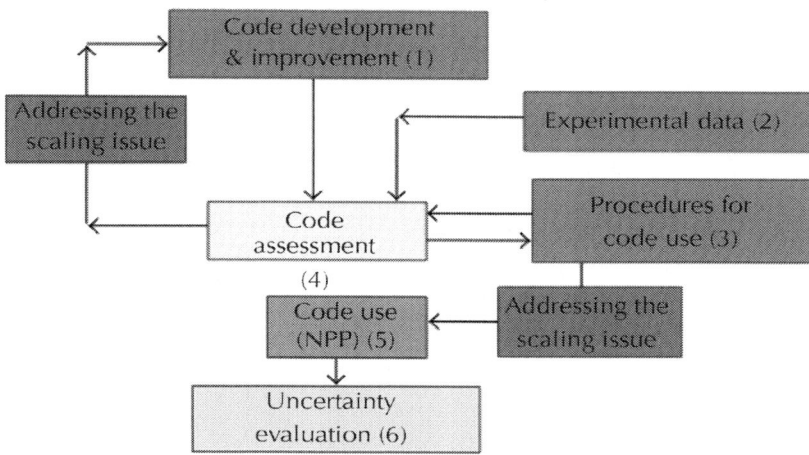

Figure 6: Role of the scaling analysis in the code assessment process.

An NPP is characterized by high power (up to thousands of MW), high pressure (tens of MPa), and large geometry (hundreds of m^3), thus it is well understandable the impossibility to perform experiments preserving all these three quantities. The term scaling is in general understood in a broad sense covering all differences existing between a real full size plant and a corresponding experimental facility. An experimental facility may be characterized by geometrical dimension and shape, arrangements, and availability of components, or by the mode of operation (e.g., nuclear versus electrical heating). All these differences have the potential to distort an experimental observation precluding its direct application for the design or operation of the reference plant. Distortion can be defined as a partial or total suppression of physical phenomena caused by only changing the size (geometric dimension) or the shape (arrangement of components) of the facility [19].

Three main objectives can be associated to the scaling analysis as follows:

- the design of a test facility;
- the code validation, that is, the demonstration that the code accuracy is scale independent;
- the extrapolation of experimental data (obtained into an ITF) to predict the NPP behavior.

For the test facility design, three types of scaling principles can be adopted as follows.

- Time-reducing scaling: rigorous reduction of any linear dimension of the test rig would result in a direct proportional reduction in time scaling. This is considered to be of advantage only for cases where body forces due to gravity acceleration are negligible compared to the local pressure differentials.
- Time preserving scale: based on a scale reduction of the volume of the loop system combined with a direct proportional scaling of energy sources and sinks (keeping constant the core power to system volume ratio).
- Idealized time preserving modeling procedures: based on the equivalency of the mathematical representation of the full size plant and of the test rig. It is deduced from a separated treatment of the conservation equations for all involved volume modes and

flow paths assuming homogeneous fluid.

Integral test facilities are normally designed to preserve geometrical similarity with the reference reactor system. Generally all main components (e.g., rector pressure vessel, downcomer, rod bundle, loop piping, etc.) and the engineered safety system (HPIS, LPIS, accumulators, auxiliary feed water, etc.) are represented. ITF are used to investigate, by direct simulation, the behavior of an NPP in case of off-normal or accident conditions. The geometrical similarity of the hardware of the loop systems has been abandoned in favor of a preservation of geometric elevations, which are decisive parameters for gravity dominated scenarios (e.g., in case of natural circulation processes). Thus the reduction of the primary system volume is largely achieved by an equivalent reduction in vertical flow cross sections.

Due to the impossibility to perform relevant experiment at full scale (i.e., in an NPP), the use of ITF or SETF is unavoidable. In order to address the scaling issue, different approaches have been proposed and are available from literature. However, a comprehensive solution has not yet been achieved and moreover, it is evident that the attempt to scale up all thermal-hydraulic phenomena that occur during an assigned transient results in a myriad of factors which have counterfeiting values [20]. For instance, let us consider Figure 7 that schematically reproduces a two-phase flow condition (TPFC) in a vessel of a facility when an SBLOCA scenario is postulated. The two-phase critical flow is affected by phenomena like the vapor pull through and the subcooled vapor formation by the sharp edge cavitations, the heat losses, the fluid temperature stratification, and so forth. All these phenomena cannot be scaled up and are characterized by parameters that do appear neither in any balance equations nor in any scalable mechanistic models. This is a typical situation in which a scaling criterion is not applicable. Nevertheless the influence of those phenomena is time-restricted in relation to the entire transient and thus they can be considered as local phenomena.

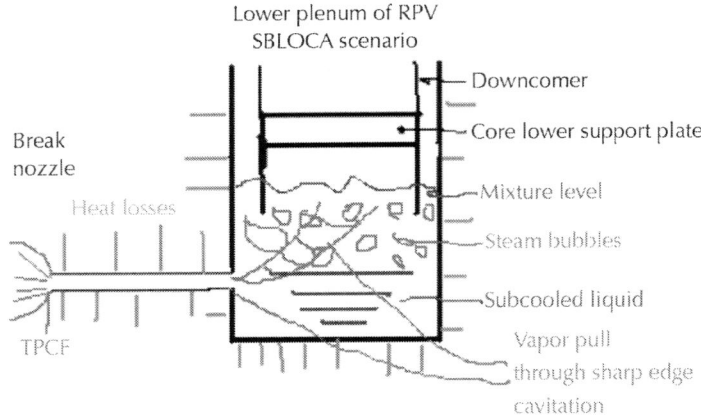

Figure 7: Schematic representation of a two-phase flow condition in a reactor pressure vessel of a facility during an SBLOCA.

As a consequence, the only way to solve the scaling problem is to consider only those phenomena and parameters that have a real impact on the whole problem under investigation. The focusing on a single phenomenon which occurs during a limited time (compared with the entire duration of the problem) should be avoided because it is governed by factors that are not scalable. Therefore a hierarchy in the definition of the scaling factors is necessary and a global strategy is needed [21] to demonstrate that those phenomena are effectively local and cannot affect the overall behavior of the main thermal-hydraulic parameters selected to describe the transient. Based on the flow diagram in Figure 6, the strategy to adopt for solving the scaling problem consists in

- developing a system code;
- qualifying the code against experimental data;
- demonstrating that the code-accuracy (i.e., discrepancy between measured and calculated trends) only depends upon boundary initial conditions (BIC) values (within the assigned variation ranges) and is not affected by the scale of concerned ITF;
- applying such code to predict the same relevant phenomena that are expected to find in a same experiment (or transient) performed at different scale;

- performing NPP Kv-scaled calculation and explaining the discrepancies (if any) between NPP Kv-scaled calculation and measured trends in ITF considering only BIC values and hardware differences (i.e., distortions).

Nodalization Qualification

Assuming the availability of a qualified code and of a qualified user, it is necessary to define a procedure to qualify the nodalization in order to obtain qualified (i.e., reliable) calculation results. In this section a procedure for the nodalization qualification is discussed.

A major issue in the use of mathematical models is constituted by the model capability to reproduce the plant or facility behavior under steady-state and transient conditions. These aspects constitute two main checks for which acceptability criteria have to be defined and satisfied during the nodalization-qualification process. The first of them is related to the geometrical fidelity of the nodalization of the reference plant; the second one is related to the capability of the code nodalization to reproduce the expected transient scenario.

The checks about the nodalization are necessary to take into account the effect of many different sources of approximations, like the following.

- The data of the reference plant available to the user are typically non exhaustive to reproduce a perfect "schematization" of the reference plant.
- From the available data, the user derives an approximated nodalization of the plant reducing the level of detail.
- The code capability to reproduce the hardware, the plant systems and the actuation logic of the systems reduce further the level of detail of the nodalization.

The reasons for the checks about the capability of the code nodalization to perform the transient analysis deriving from following considerations:

- the code options must be adequate;
- the nodalization solutions must be adequate;

- some systems components can be tested only during transient conditions (e.g., ECCS that are not involved in the normal operation).A simplified scheme of a procedure that can be adopted for the qualification of the nodalization is depicted in Figure 8 [22]. In the following, it has been assumed that the code has fulfilled the validation and qualification process and a "frozen" version of the code has been made available to the final user. This means that the code user does not have the possibility to modify or change the physical and numerical models of the code (only the options described in the user manual are available to the user). With reference to Figure 8, the qualification procedure of the nodalization is described step by step.

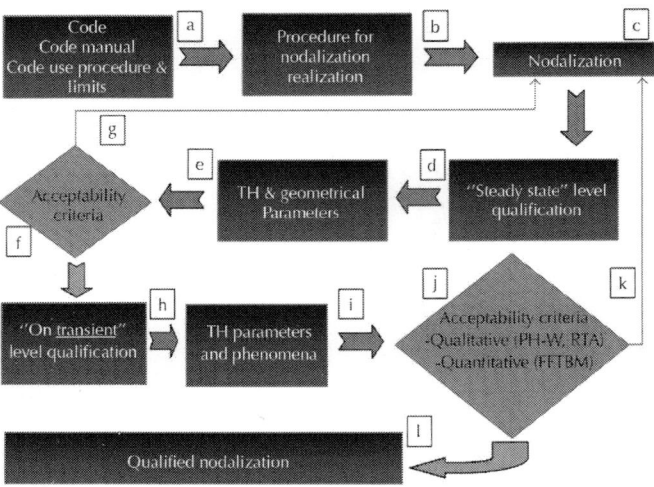

Figure 8: Flow sheet of nodalization qualification procedure.

Step "a"

This step is related to the information available by the user manual and by the guidelines for the use of the code. This type of information takes into account the specific limits and assumptions of the code (specific of the code adopted for the analysis) and some guidelines deriving from the best practices for realizing the nodalization. From a generic point of view, the following aspects should be carefully adopted:

- homogeneous nodalizations;
- strict observation of the user guidelines;
- standard use of the code options.

Step "b"

User experience and developers recommendations are useful to set up particular procedure to be applied for a better nodalization. These special procedures are related to the specific code adopted for the analysis. An example is constituted by the "slice nodalization" technique adopted with the RELAP5 code to improve the capability of the code to reproduce transients involving natural circulation phenomena.

Step "c"

The realization of the nodalization depends on several aspects: available data, user capability and experience, code capability. The nodalization must reproduce all the relevant parts of the reference plant; this includes geometrical and materials fidelity and reproduction of the systems and related logics. From a generic point of view, the following recommendations can be done.

- Data must be qualified or in other words, data has to derive from
- qualified data facility (if the analysis is performed for a facility);
- qualified test design;
- qualified test data.

1. The data base for the realization of the nodalization should be derived from official document and traceability of each reference should be maintained. However three different types of data can be identified as follows:

- qualified data, from official sources;
- data deriving from nonofficial sources; these types of data can be derived from similar plant data, or other qualified nodalization for the same type of plant; the use of these data can introduces potential errors and the effect on the calculation results must be carefully evaluated;
- data assumed by the user; these data constitute some assumptions of the user (on the base of the experience or by similitude with

other similar plants). The use of this type of data should be avoided. Any special assumptions adopted by the user or special solutions in the nodalization must be recorded and documented.

Step "d"

The "steady-state" qualification level includes different checks: one is related to the evaluation of the geometrical data and of numerical values implemented in the nodalization; the other one is related to the capability of the nodalization to reproduce the steady-state qualified conditions. The first check should be performed by a user different from the user has developed the nodalization. In the second check a "steady-state" calculation is performed. This activity depends on the different code peculiarities. As an example, for RELAP5, the steady-state calculation is constituted by a "null-transient" calculation (i.e., the "transient" option is selected and no variation of relevant parameters occurs during the calculation).

Step "e"

The relevant geometrical values and the relevant thermal-hydraulic parameters of the steady-state conditions are identified. The selected geometrical values and the selected relevant parameters are derived, respectively, from the input deck of the nodalization and from the steady-state calculation for performing the comparison with the hardware values and the experimental parameters.

Step "f"

This is the step where the adopted acceptability criteria are applied to evaluate the comparison between hardware and implemented geometrical values in the nodalization (e.g., volumes, heat transfer area, etc.) and between the experimental and calculated steady-state parameters (e.g., pressures, temperatures, mass flow rates, etc.). Some comments can be added as follows.

- The experimental data are typically available with error bands which must be considered in the comparison with the calculated values and parameters.

- The steadiness of the steady-state calculation must be checked.

Step "g"

If one or more than one of the checks in the step "f" are not fulfilled, a review of the nodalization (step "c") must be performed. This process can request more detailed data, improvement in the development of the nodalization, different user choices. The path "g" must be repeated till all acceptability criteria are satisfied. A list of the geometrical values and of the thermal-hydraulic parameters to be checked is given in Table 1together with acceptable errors.

Table 1: Parameters and acceptable errors for the nodalization qualification at "steady-state" level

	Quantity	Acceptable error (°)
1	Primary circuit volume	1%
2	Secondary circuit volume	2%
3	Nonactive structure heat transfer area (overall)	10%
4	Active structure heat transfer area (overall)	0.1%
5	Non-active structure heat transfer volume (overall)	14%
6	Active structure heat transfer volume (overall)	0.2%
7	Volume versus height curve (i.e., "local" primary and secondary circuit volume)	10%
8	Component relative elevation	0.01 m
9	Axial and radial power distribution (° °)	1%
10	Flow area of components like valves, pumps orifices	1%
11	Generic flow area	10%

(*)		
12	Primary circuit power balance	2%
13	Secondary circuit power balance	2%
14	Absolute pressure (PRZ, SG, ACC)	0.1%
15	Fluid temperature	0.5% (**)
16	Rod surface temperature	10 K
17	Pump velocity	1%
18	Heat losses	10%
19	Local pressure drops	10% (^)
20	Mass inventory in primary circuit	2% (^^)
21	Mass inventory in secondary circuit	5% (^^)
22	Flow rates (primary and secondary circuit)	2%
23	Bypass mass flow rates	10%
24	Pressurizer level (collapsed)	0.05 m
25	Secondary side or downcomer level	0.1 m (^^)

°The % error is defined as the ratio (reference or measured value—calculated value). The "dimensional error" is the numerator of the above expression. *With reference to each of the quantities below, following a one-hundred-second "null-transient" calculation, the solution must be stable with an inherent drift <1%/100 second. **And consistent with power error. ^Of the difference between the maximum and minimum pressure in the loop. ^^And consistent with other errors.

Step "h"

This step constitutes the "On Transient" level qualification. This activity is necessary to demonstrate the capability of the code nodalization to reproduce the relevant thermal-hydraulic phenomena expected during the transient. This step also permits to verify the correctness of some systems that are in operation only during transient events. Criteria, both qualitative and quantitative, are established to express the acceptability of the transient calculation. Two different aspects can be identified as follows.

- The code input deck concerns with the nodalization of an ITF. In this case the code calculation is used for the code assessment.

Checks include the code options selected by the user, the solutions adopted for the development of the ITF nodalization, the logic of some systems (e.g., ECCS). Typically many experimental results are available, thus a similar test can be adopted for performing the "On Transient" level qualification.

- The objective of the code calculation is constituted by the analysis of a transient in an NPP. In this case, it is necessary to check the nodalization capability to reproduce the expected thermal-hydraulic phenomena occurring during the transient, the selected code options, the adopted solutions for the development of the NPP nodalization, and the logic of the systems not involved in the steady-state calculation. Typically no data exist for the transients performed in the NPP. For this reason, data from experiments carried out in ITF can be used for performing the so-called "Kv-scaled" calculation. The Kv-scaled calculation consists in using the developed NPP nodalization for predicting an experimental transient (whose kind is similar to the one under investigation in the NPP) performed in an ITF. The NPP nodalization is prepared for the Kv-scaled calculation by properly scaling the BICs characterizing the selected transient in the ITF. In other words, power, mass flow rates and ECCS capacity are scaled adopting as scaling factor the ratio between the volume of the facility and the volume of the NPP. The capability of the nodalization to reproduce the same transient evolution and the thermal-hydraulic relevant phenomena is the needed request for satisfying the "On Transient" qualification level.

Step "i"

In this step the relevant thermal-hydraulic phenomena and parameters are selected and a comparison between the calculated and experimental data is performed. The selection of the phenomena derives from the following sources:

- experimental data analysis (engineering judgment is request);
- CSNI phenomena identification;
- use of Relevant Thermal-hydraulic Aspects (RTA, engineering judgment is request).

Step "j"

This is the step where checks are performed to evaluate the acceptability of the calculation both from qualitative and from quantitative point of view. For the qualitative evaluation the following aspects are involved:

- Visual observation. This means that a visual comparison is performed between experimental and calculated relevant parameters time trends;

- Sequence of the resulting events. This means that the list of the calculated significant events together with their timing of occurrence is compared with the experimental events;

- Use of the CSNI phenomena. The relevant phenomena suitable for the code assessment and their relevance in the selected facility and in the selected test are identified. A judgment can be express taking into account the characteristics of the facility, the test peculiarities and the code results;

- Use of the RTAs. RTAs are typically identified inside the phenomenological windows (i.e., time windows where a unique relevant phenomenon is occurring) and are characterized by special parameters. These parameters can be time values, single values, integral values, gradient values and nondimensional values. An example of a table containing RTAs is given in Table 2.Quantitative checks are carried out by using the Fast Fourier Transform Based Method (FFTBM). This special tool performs the comparison between experimental and calculated time trends in the frequency domain for a list of selected parameters and calculates, for each of them, a numerical value by which the accuracy is quantitatively evaluated (no engineering judgment is involved in this process). The FFTBM makes also possible to obtain a numerical judgment of the overall results of the calculation. Criteria based on the values attained by FFTBM had been selected for accepting the transient calculation. A description of the FFTBM can be found in [23].

Table 2

		UNIT	EXP	UNIPI 91BN1OLPSI	CEA c2m4 Icea	Judgment UNIPI/CEA
RTA: pressurizer emptying						
TSE	Emptying time*	s	131	46	—	R/-
	Scram time	s	41	38	41	R/E
RTA: steam generators secondary side behaviour						
TSE	Main feed water off, turbine bypass	s	59	55	42	E/R
SVP	Difference between PS and SG 1 SS pressure at 100 s	MPa	0.42	0.33	0.37	R/R
SVP	SG 1 mass	Kg/(s)				
	at the end of subcooled blowdown		774/(82)	781/(75)	761/(82)	E/E
	when PS pressure equals SG 1 SS pressure		869/(618)	938/(408)	847/(463)	R/R
	when ACC starts		804/(2955)	802/(3019)	788/(3075)	E/R
	when LPIS starts		938/(5176)	1126/(6529)	956/(5474)	R/R
SYP	SG 1 pressure	MPa				
	at the end of subcooled blowdown		7.15	7.10	7.05	E/E
	when PS pressure equals SS pressure		6.95	7.04	7.03	R/R
	when ACC starts		4.11	3.95	4.00	R/E
	when LPIS starts		0.88	0.83	0.83	E/E
RTA: subcooled blowdown						

TSE	Upper plenum in sat conditions	s	83	100	110	R/R
IPA	Break flow up to 100 s	kg	152	161	162	R/R
RTA: first dryout occurrence						
TSE	Time of dryout	s	2237	2299	2444	E/R
	Range of dryout occurrence at various core levels	s	2237÷2471	2299÷2518	2444÷2625	R/R

Step "k"

This path is actuated if any of the checks (qualitative and quantitative) is not fulfilled. The nodalization is improved by adopting different noding solutions, changing code options or increasing the level of detail using, if available, more precise data. Every time the nodalization is modified a new qualification process will be performed through the loop "c-d-e-f-h-i-j-c."

Step "l"

This is the last step of the procedure. The obtained nodalization is used for the selected transient and the selected facility or plant. Any subsequent modification of the nodalization (e.g., necessary to better reproduce the experimental results) requires a new qualification process both at "steady-state" and "on transient" level.

DEVELOPMENT AND USE OF COUPLED COMPUTER CODES

Complex computer codes are used for the analysis of the performance of NPPs. They include many types of codes that can be grouped in different categories [24] like reactor physics codes; fuel behavior codes; thermal-hydraulic codes, including system codes, subchannel codes, porous media codes and computational fluid dynamic (CFD) codes; containment analysis codes; atmospheric dispersion and dose codes and structural codes.

Historically, these codes have been developed independently, but have been mainly used in combination with system thermal-hydraulic codes. By increasing the capacity of computation technology, safety experts thought of coupling these codes in order to reduce uncertainties or errors associated with the transfer of interface data and to improve the accuracy of calculation. The coupling of primary system thermal-hydraulics with neutronics is a typical example of code coupling; other cases include coupling of primary system thermal-hydraulics with structural mechanics, fission product chemistry, computational fluid dynamics, nuclear fuel behavior and containment behavior. Problems that need to be addressed in the development and use of coupled codes include ensuring adequate computer capacity and efficient coupling procedures, validation of coupled codes and evaluation of uncertainties, and consequently the applicability of coupled codes for safety analyses.

The major purposes of the development of coupled code are to be capable of representing the results of interactions between different physical phenomena in more detail. Since the calculation method of each code is not changed, reduction of computational time or necessary computer memory volume is not expected. Nevertheless, many additive benefits are expected as follows.

- Since the interface data are easily, automatically and frequently exchanged between codes, the results of calculation would be obtained faster than the combination of individual codes and also be more reliable.

- Since the development works are limited to the interface part, the cost and time for development can be minimized.

- Since the interface data between each code would be adjusted to meet the specifications (e.g., noding of the system or time increment of calculation) of each code at the development stage, additional assumptions or data averaging and reductions are not required when performing the calculation.

- Those that have the knowledge of the existing codes are not necessary to study the coupled code from the beginning, because the existing knowledge is applicable to the coupled code.It is expected that those benefits can contribute to the improvement of activities carried out by both licensing authorities and industries. Expectations for licensing authorities can mainly be derived from the features of coupled codes such as more accurate calculation than the combination of individual codes. These are summarized as follows:

a.   improvement of the understanding of the phenomena of interest for safety;

b.   better assessment/demonstration of the conservatisms (versus historical approaches such as the use of point kinetics or evaluation models);

c.   extension of the capabilities of the codes for safety analysis and training/simulators;

d.   better assessment of uncertainties associated with the use of best estimate couplet codes.

Many benefits are expected with the use of coupled codes for industries. These are as follows.

a.   Faster turnaround of calculation allows the users to perform more precise analysis and more sensitivity or case studies. This would contribute in more detail to understand the features of the plant, systems or components.

b.   More accurate calculation would contribute to remove unnecessary uncertainties and to identify margins available to use for the plant.

c.   Uncertainties due to user effects would be minimized because the existing knowledge of individual codes is applicable to the coupled codes.

The request to use qualified tools in licensing calculations constitutes one of the main problems to be addressed in the development of coupled computer codes and it is caused by the limited availability of data, which can be obtained from operating plants. To reduce the effort for the qualification of the coupled codes, code developers are requested to use only validated revisions of codes. In addition, the code developers are requested to

a.   design the coupling so that auditing is easy and feasible;

b.   provide guidelines to minimize user effects;

c.   allow provisions for reasonable conservatisms;

d.   structure the code so that coupling is easy and feasible;
e.   standardize the coupling procedures;
f.   integrate as much as possible the existing approved calculation methodologies.

CONCLUSIONS

A noticeable progress in the capabilities of system codes has been observed in the past decades. From the design and safety engineering point of view, thermal-hydraulic system codes are considered to have reached an acceptable level of maturity. Most of the problems and questions that come up a couple of decades ago have been solved or an answer has been proposed. In other words, there is more need to synthesize the work done in the international ground than to identify new problems. For instance, if corresponding measured and calculated trends are given, possible research should be focused on answering whether the discrepancy is acceptable and less on minimizing the discrepancy itself (e.g., through an improved model). It is evident that all the progress has been made in the recent past is a consequence of experimental researches. After 30 years of validation through basic, separate and integral effect tests facilities and after code improvements, system codes are able to predict main phenomena of PWR & BWR transients with reasonable accuracy. Nowadays the attention should be focused more on developing procedures for a consistent application of a thermal-hydraulic system code. This need has been highlighted in the paper and implies the drawing up of specific criteria through which the code-user, the nodalization and finally the calculated transient results can be qualified.

The full exploitation of "advanced" best-estimate system codes (e.g., TRAC, RELAP, ATHLET, CATHARE), which are strictly based on two-fluid representation of two-phase flow and a "best-estimate" description (in contrast with the evaluation models which used many conservative assumptions) of complex flow and heat transfer conditions, implies mainly their acceptability by the licensing authorities. In fact, notwithstanding the important achievements and progresses made in the recent years, the predictions of advanced best-estimate computer codes are not exact but remain uncertain because of the following.

- The assessment process depends upon data almost always measured in small scaled facilities and not in the full power reactors.
- (The models and the solution methods in the codes are approximate: in some cases, fundamental laws of the physics are not considered.

Consequently, the results of the best estimate code calculations may not be applicable to give "exact" information on the behavior of an NPP during postulated accident scenarios. Therefore, best-estimate analysis must be supplemented by proper uncertainty evaluations in order to be meaningful and conditions for their application should be made clear for accepting the available uncertainty methods in the licensing process.

In conclusion, the present status, of system codes development, assessment, and related uncertainty evaluation, is adequate as far as the largest majority of design and safety problems of current water-cooled reactors are concerned. Anyway, new scientific goals must be achieved. To this aim, projects and programmes based on the development of system codes with multidimensional and multifluid capability and with "open" interfaces for an easy coupling with other codes in areas like neutronics (for implementing presently available 3D codes), CFD, structural mechanics (e.g., for pressurized thermal-shock studies), and containment constitute the new frontier of the scientific and engineering community in this field. However, taking into account that the development of such codes with measurable increased improvements in their capabilities may need several decades, it is an evident consequence that the existing system thermal-hydraulic codes are going to be used for one or two decades in their present configuration.

REFERENCES

1.   "Acceptance criteria for emergency core cooling systems (ECCS) in light water nuclear reactors (10CFR 50.46)," Appendix K to Part 50 "ECCS Evaluation Models", Federal Register, vol. 43, no. 235 (43 FR 57157), December, 197.

2. S. Belsito, F. D'Auria, and G. M. Galassi, "Application of a statistical model to the evaluation of counterpart test database," Kerntechnik, vol. 59, no. 3, 1994.

3. N. Aksan, N. D'Auria, H. Glaeser, R. Pochard, C. Richards, and A. Sjoberg, "A separate effects test matrix for thermal-hydraulic code validation: phenomena characterization and selection of facilities and tests," OECD/GD (94) 82, vols. I and II, 199.

4. N. Aksan, D. Bessette, I. Brittain, et al., "Code validation matrix of thermo-hydraulic codes for LWR LOCA and transients," CSNI Report 132, March 1987.

5. M. J. Lewis", , R. Pochard, F. D'Auria, et al., "Thermohydraulics of emergency core cooling in light water reactors-a state of the art report," OECD/CSNI 161, October 198.

6. USNRC, "Compendium of ECCS research for realistic LOCA analysis," NUREG-1230, December 198.

7. B. E. Boyack, I. Catton, R. B. Duffey, et al., "Quantifying reactor safety margins—part 1: an overview of the code scaling, applicability, and uncertainty evaluation methodology," Nuclear Engineering and Design, vol. 119, no. 1, pp. 1–15, 1990.

8. N. Aksan, F. D'Auria, H. Glaeser, R. Pochard, C. Richards, and A. Sjoberg, "Overview of the CSNI separate effects test validation matrix," in Proceedings of the 7th International Topical Meeting on Reactor Thermal Hydraulics (NURETH '95), New York, NY, USA, September 1995.

9. S. N. Aksan, F. D'Auria, and H. Städtke, "User effects on the thermal-hydraulic transient system code calculations," Nuclear Engineering and Design, vol. 145, no. 1-2, pp. 159–174, 1993.

10. M. Bonuccelli, F. D'Auria, N. Debrecin, and G. M. Galassi, "A methodology for the qualification of thermalhydraulic codes nodalizations," in Proceedings of the 5th International Topical Meeting on Reactor Thermal Hydraulics (NURETH '93), Grenoble, France, October 1993.

11. F. D'Auria, M. Leonardi, and R. Pochard, "Methodology for the evaluation of thermalhydraulic codes accuracy," in Proceedings of International Conference on New Trends in Nuclear System Thermalhydraulics, Pisa, Italy, May-June 1994.

12.   F. D'Auria and W. Giannotti, "Development of code with capability of internal assessment of uncertainty," Nuclear Technology, vol. 131, no. 1, pp. 159–196, 2000.

13.   A. Petruzzi, F. D'Auria, W. Giannotti, and K. Ivanov, "Methodology of internal assessment of uncertainty and extension to neutron kinetics/thermal-hydraulics coupled codes," Nuclear Science and Engineering, vol. 149, no. 2, pp. 211–236, 2005.

14.   T. Wickett, et al., "Report of the uncertainty method study for advanced best estimate thermal-hydraulic code applications," 1998, vols. I and II, OECD/CSNI Report NEA/CSNI R (97) 35, Paris, Franc.

15.   A. Petruzzi, et al., "BEMUSE Programme. Phase 2 report (re-analysis of the ISP-13 exercise, post test analysis of the LOFT L2-5 experiment)," 2006, OECD/CSNI Report NEA/CSNI/R(2006)2, pp. 1–62.A. De Crecy, et al., "BEMUSE Programme. Phase 3 report (uncertainty and sensitivity analysis of the LOFT L2-5 experiment)," 2007, OECD/CSNI Report NEA/CSNI/R(2007).F. D'Auria and G. M. Galassi, "Code validation and uncertainties in system thermalhydraulics," Progress in Nuclear Energy, vol. 33, no. 1-2, pp. 175–216, 1998.

16.   N. Aksan, "International standard problems and small break loss-of-coolant accident (SBLOCA)," inProceedings of THICKET: Seminar on Transfer of Competence, Knowledge and Experience Gained Through CSNI Activities in the Field of Thermalhydraulics, CSNI OECD/NEA, INSTN and IRSN, Saclay, France, June 2004.

17.   F. D'Auria, K. Fischer, B. Mavko, and A. Sartmandjiev, "Validation of accident and safety analysis methodology," Internal Technical Report, International Atomic Energy Agency, Vienna, Austria, June 2001.

18.   N. Zuber, G. E. Wilson, M. Ishii, et al., "An integrated structure and scaling methodology for severe accident technical issue resolution: development of methodology," Nuclear Engineering and Design, vol. 186, no. 1-2, pp. 1–21, 1998.

19.   F. D'Auria, G. M. Galassi, and P. Gatta, "Scaling in nuclear system thermal hydraulics: a way to utilise the available database," in Proceedings of the 32nd National Heat Transfer Conference (ASME '97), vol. 350, pp. 35–43, Baltimore, Md, USA, 1997.

20. A. Petruzzi, F. D'Auria, and W. Giannotti, "Description of the procedure to qualify the nodalization and to analyze the code results," DIMNP NT 557(05), May, 200.

21. A. Petruzzi and F. D'Auria, "Accuracy quantification: description of the fast fourier transform based method (FFTBM)," DIMNP NT 556(05), May, 200.

22. IAEA-TECDOC-1539 (2007), "Use and development of couplet computer codes for the analysis of accidents at nuclear power plants," in International Atomic Energy Agency, Technical Meeting, Vienna, Austria, November 2003.

Chapter 3

Finite Element Modeling of Shop Built Spherical Pressure Vessels

Oludele Adeyefa and Oluleke Oluwole

Department of Mechanical Engineering, University of Ibadan, Ibadan, Nigeria

ABSTRACT

This work builds on an earlier work done which used global coordinates where a large number of elements were needed to form a convergence of results for shop built spherical pressure vessels. In this work area coordinates were used. Any action that leads to an inability on the part of a structure to function as intended is known as failure. This research, therefore, investigates stresses developed in a shop built carbon steel spherical storage vessels using finite element approach as the analytical tool. 3-D finite element modeling using 3-node shallow triangular element with five degrees of freedom at each node is employed. These five degrees of freedom are the essential nodal degrees of freedom without the sixth in-plane rotation. The resulting equations from finite element analysis are coded using FORTRAN 90 computer programme. Spherical storage vessels are subjected to various internal loading pressures while nodal displacements, strains and the corresponding maximum Von-mises stresses are determined. The calculated maximum Vonmises stresses are compared with the yield strength of the shell plate material. Using specified safety factor, safety internal pressures with the corresponding shell thicknesses for shop built spherical pressure vessels are determined. The finite element modeling carried out in this research can be used to predict in-service stresses, strains, and deformations of shop built spherical pressure vessels using Von-mises yield stress as the failure criteria. The results obtained were validated by analytical method and it showed there was no significant difference (P > 0.05) with values obtained through analytical method.

INTRODUCTION

Considerable attention has been given to applying the finite element method in the analysis of curved structures. [1] developed conical segments for the analysis of shell of revolution. [2] modified the method and used meridional elements which were found to lead to considerably improved results for the stresses. Curved rectangular and cylindrical shell elements were also developed [3-5]. Applications have been in the area of membrane, thermal and pressure analysis [6-11].

However, to model a shell of spherical shape using the finite element method triangular and rectangular spherical shell elements are needed. [12] used curved shallow triangular element in curvilinear coordinates system to model spherical storage pressure vessels. The assumed polynomial function for the bending behaviour in their model is in term of x and y without the term x^2y while the function for membrane behaviour is linear. The 3- node shallow triangular element has five degrees of freedom at each node without the in-plane rotation.

This approach gives considerable and acceptable results but at the expense of many elements leading to higher storage and computational efforts. It is observed that exclusion of tenth term in the polynomial function for the assumed polynomial represent deformation pattern for the bending of shell element accounted for the requirement for more elements. Meanwhile, to analyse the plate bending behaviour using triangular elements, in [13] classified such element as non-conforming element due to the incomplete polynomial terms in representing bending deformation pattern. [13] proposed polynomial function for the bending behaviour to be represented using area coordinated. It is said of this element derived in [13] that it passes all the patch tests and performs excellently. Indeed if the quadrature is carried out in a "reduced manner" using three quadrature points, then the element is one of the best triangles with 9 degrees of freedom that is currently available in [13]. This approach is being employed in this work for the FE modelling of spherical shell. The membrane deformation of shallow triangular shell element is represented with polynomial linear function using area coordinates. The bending deformation is assumed as it is given by in [13] for plate bending using 9 degree of freedom using area coordinates for non-conforming elements.

FINITE ELEMENT METHODOLOGY

A detailed study of stress analysis of shop built spherical pressure vessel subjected to different internal and external pressures is carried out with the help of finite element method, which is perhaps the best currently known method available for the stress analysis of pressure vessel problems.

Displacement Field Requirements

The assumed displacement method was employed in this work to develop a shallow triangular spherical shell element without an in-plane rotation as a sixth degree of freedom. A shallow shell formulation was used to obtain the displacement fields. The element (Figure 1) has five degrees of freedom at each of the three corners [12]. Therefore, there are fifteen degrees of freedom per element. However, the assumed deformation pattern for bending of the shallow triangular shell element was according to the one given by Zienkiewicz et al. [13] in area coordinates while the deformation for the membrane is represented by linear polynomial using area coordinates as opposed to an earlier work that used global coordinates [12].

Basic Assumptions of the Analysis

The spherical shallow shell under investigation is assumed to have the following properties and to be loaded in the following manner:

- The spherical shell is taken as thin shell.
- The loads to be considered are internal pressure and external pressure.
- The displacements of the vessel are assumed to be so small that the equilibrium conditions for an element in the spherical shell is the same before and after deformation.

Displacement Functions

The assumed displacement relationships for the proposed triangular shallow shell are expressed in curvilinear coordinates in area coordinates. The use of so-called "area coordinates" is made to represent the transverse displacement, w, as a polynomial function of degree 3 as it was given by [13]. Linear polynomial equations are then used to represent the membrane displacements u and v using area coordinates, resulting in a constant strain triangle for the membrane action. The assumed displacement equations are:

$$u = a_1 L_1 + a_2 L_2 + a_3 L_3 \tag{1}$$

$$u = a_4 L_1 + a_5 L_2 + a_6 L_3 \tag{2}$$

$$
\begin{aligned}
w = {} & a_7 L_1 + a_8 L_2 + a_9 L_3 + a_{10} L_1 L_2 + a_{11} L_2 L_3 + a_{12} L_3 L_1 \\
& + a_{13}\left[L_1^2 L_2 + \frac{1}{2} L_1 L_2 L_3 \left\{ 3(1-\mu_3) L_1 - (1-3\mu_3) L_2 + (1-3\mu_3) L_3 \right\} \right] \\
& + a_{14}\left[L_1^2 L_2 + \frac{1}{2} L_1 L_2 L_3 \left\{ 3(1-\mu_3) L_1 - (1-3\mu_3) L_2 + (1-3\mu_3) L_3 \right\} \right] \\
& + a_{15}\left[L_1^2 L_2 + \frac{1}{2} L_1 L_2 L_3 \left\{ 3(1-\mu_3) L_1 - (1-3\mu_3) L_2 + (1-3\mu_3) L_3 \right\} \right]
\end{aligned}
\tag{3}
$$

$$\theta_x = \frac{\partial w}{\partial y} \tag{4}$$

$$\theta_y = -\frac{\partial w}{\partial x} \tag{5}$$

Where

$$\mu_i = \frac{l_k^2 - l_j^2}{l_i^2}$$

and l_i is the length of the side opposite node i. The modified interpolation for displacement is taken as

$$\phi = Pa \tag{6}$$

to determine constants as, known displacements at nodes are substituted and the equations become

$$[a] = \left[C^{-1} \right][\delta] \tag{7}$$

where $[\delta]$ is the nodal degrees of freedom, $[C^{-1}]$ is inverse of transformation matrix and $[a]$ is vector of independent constants.

Strain-Displacement Equations

Strain-displacement relationships for shallow thin shells as given by [14] are simplified for the shallow shell and expressed as follows in curvilinear coordinates.

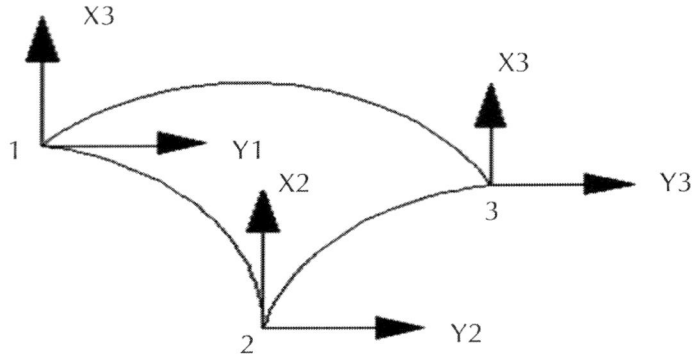

Figure 1: Shallow triangular element.

$$\varepsilon_x = \frac{\partial u}{\partial x} + \frac{w}{r}, \quad \varepsilon_y = \frac{\partial v}{\partial y} + \frac{w}{r}, \quad \varepsilon_y = \frac{\partial v}{\partial y} + \frac{w}{r},$$

$$k_x = -\frac{\partial^2 w}{\partial x^2}, \quad k_{xy} = -2\frac{\partial^2 w}{\partial x \partial y}$$

(8)

The above strain Equation (8) which can be written in matrix form after necessary substitutions of u, v and w given in Equations (1)-(3) into the above strain equations.

Stress in a Curved Triangular Element

Stress varies from point to point along the shell profile and also through the thickness of the shell. It is thus in reality an unknown function of two variables, therefore leads us to the equations below as it were given by [15]:

$$\sigma_b = \frac{6M}{t^2}, \quad \sigma_m = \frac{N}{t}$$

(9)

where: M is the moment per unit length, M and σ_b is the bending stress at the surface.

N is to be force per unit length and σ_m which is membrane stress.

Strain Energy

The strain energy equation for an isotropic linear shell is

$$U =$$

$$\iint_A \int_{-\frac{t}{2}}^{\frac{t}{2}} \frac{E}{2(1-v^2)} \left[\varepsilon_x^2 + \varepsilon_y^2 + 2v\varepsilon_x\varepsilon_y + \frac{1}{2}(1-v)\gamma_{xy}^2 \right] d\varsigma dx dy \qquad (10)$$

where, t = thickness of the shell, v = Poisson's ratio and E = Modulus of elasticity.

After substitution for strains in the above expression and integration with respect to, the strain energy can be separated into the membrane energy U_m and the bending energy U_b.

$$U = U_m + U_b \qquad (11)$$

$$U_m = \frac{Et}{2(1-v^2)} \iint_A \left[e_x^2 + e_y^2 + 2ve_xe_y + \frac{1}{2}(1-v)e_{xy}^2 \right] dxdy \qquad (12)$$

$$U_b = \frac{Et^3}{24(1-v^2)} \iint_A \left[k_x^2 + k_y^2 + 2vk_xk_y + \frac{1}{2}(1-v)k_{xy}^2 \right] dxdy \qquad (13)$$

The potential energy is then written as: $\Phi = U - W$ where W represents the work done by the external load on the system. In the finite element method, the potential energy of a shell is expressed as:

$$\Phi = \sum_{k=1}^{n} \phi_k \qquad (14)$$

where ϕ_k is the potential energy of the kth element.

Stiffness Matrix

The stiffness matrix of the elements is derived from the principle of minimum potential energy, using the theory of shallow shells, which is more accurate enough for the shallow curved triangular elements considered. By writing strain energy equations in terms of area coordinates, element stiffness matrix can be determined in the usual manner,

$$k_m = t\left[C^{-1}\right]^T \iint_A B_m^T D_m B \mathrm{d}A \left[C^{-1}\right]$$

(15)

$$k_b = t\left[C^{-1}\right]^T \iint_A B_b^T D_b B \mathrm{d}A \left[C^{-1}\right]$$

(16)

k_m and k_b are element stiffness matrices due to membrane and bending stresses respectively.

D_m and D_b are elasticity matrices for membrane and bending stresses respectively.

B_m and B_b are strain matrices for membrane and bending stresses respectively.

Therefore, element total stiffness matrix is

$$k = k_b + k_m$$

(17)

Element stiffness matrix is then combined to give system stiffness matrix. It is to note that stiffness matrices kb and km are in terms of area coordinates which can be integrated explicitly or in a reduced "manner" using three Gauss quadrature points. To integrate explicitly, the integral equation below as it is in [13] is very useful.

$$\int_A L_1^a L_2^b L_3^c \mathrm{d}A = \frac{a!b!c!}{(a+b+c+2)!} 2\Delta$$

(18)

where Δ is the area of triangular element.

Consistent Load Vector

It is well known fact that the best and accurate approach for dealing with distributed loads in FEM is the use of a consistent load vector which is derived by equating the work done by the distributed load through the displacement of the element to the work done by the nodal generalized loads through the nodal displacements. If a shallow triangular shell element is acted upon by a distributed load q per unit area in the direction of w, the work done by this load is given by:

$$P_1 = \int_A qw\,dxdy$$

(19)

If w is taken to be represented by:

$$\{w\} = [P]\{a\} = [P][C^{-1}]\{\delta\}$$

(20)

where [P] for the present element is given as

$$P = [0 \ \ 0 \ \ 0 \ \ 0 \ \ 0 \ \ 0 \ \ L_1 \ \ L_2 \ \ L_3 \ \ L_1 L_2 \ \ L_2 L_3 \ \ L_3 L_1 \ \ P_{13} \ \ P_{14} \ \ P_{15}]$$

(21)

Where

$$P_{13} = \left[L_1^2 L_2 + \frac{1}{2} L_1 L_2 L_3 \left\{ 3(1-\mu_3)L_1 - (1-3\mu_3)L_2 + (1-3\mu_3)L_3 \right\} \right]$$

$$P_{14} = \left[L_1^2 L_2 + \frac{1}{2} L_1 L_2 L_3 \left\{ 3(1-\mu_3)L_1 - (1-3\mu_3)L_2 + (1-3\mu_3)L_3 \right\} \right]$$

$$P_{15} = \left[L_1^2 L_2 + \frac{1}{2} L_1 L_2 L_3 \left\{ 3(1-\mu_3)L_1 - (1-3\mu_3)L_2 + (1-3\mu_3)L_3 \right\} \right]$$

and μ is as defined in Section 2.2.

The work done by the consistent nodal generalized force through the nodal displacements {δ} is given by:

$$P_2 = \{F^T\}\{\delta\}$$

(22)

Hence, from Equations (19)-(22), the nodal forces are obtained

$$F = [C^{-1}]^T \int [P^T] q\,dxdy$$

(23)

Equation (23) gives the nodal forces for a single element; and the nodal forces for the whole structure are obtained by assembling the elements' nodal forces.

Boundary Conditions

Before the system equations are ready for solution, they must be modified to account for the boundary conditions of the problem. At this junction, there is a need to give known displacement. For this system, it is assumed that displacements in all directions with the exception of radial direction are known to be zero. Also, symmetry nature of the system is taking into consideration by using 1/6th of the spherical vessel and thereby reducing the computing time. Shown in Figure 2 is a sample of the mesh with 4 elements and 6 nodes. Using area coordinates, the need for different sector angles was eliminated as was the case in [12].

PROBLEM CONSIDERED

Case One

Maximum equivalent Von-Mises stresses and factor of safety determined for a spherical vessel with the material properties and simulation conditions:

Shell material specified minimum yield stress = 240 Mpa;

Shell material allowable stress = 128 Mpa;

Shell material = A516M Grade 65;

Shell material density = 7850 kg/m³;

Spherical vessel radius = 1.0 m;

Internal pressure = 10,000 N/m²;

Factor of safety based on allowable stress = 1.875.

Case Two

Maximum equivalent Von-Mises stresses and factor of safety determined for a spherical vessel with the material properties and simulation conditions:

Shell material specified minimum yield stress = 240 Mpa;

Shell material allowable stress = 128 Mpa;

Shell material = A516 M Grade 65;

Shell material density = 7850 kg/m³;

Spherical vessel radius = 1.0 m;

Internal pressure = 20,000 N/m²;

Factor of safety based on allowable stress = 1.875.

Case Three

Maximum equivalent Von-Mises stresses and factor of safety determined for a spherical vessel with the material properties and simulation conditions:

Shell material specified minimum yield stress = 240 Mpa;

Shell material allowable stress = 128 Mpa;

Shell material = A516 M Grade 65;

Shell material density = 7850 kg/m³;

Spherical vessel radius = 0.5 m;

Internal pressure = 30,000 N/m²;

Factor of safety based on allowable stress = 1.875.

RESULTS AND DISCUSSIONS

By examining Tables 1-3, element thicknesses given in each case are uniform. This is expected because; the spherical pressure vessels in each of the cases considered was subjected to uniform internal pressure. Also, developed and Von-mises stresses calculated in each of the cases are well below the material allowable stress values as it is given by ASME sec VIII, Part D. This can be deduced from the values of factor of safety in the above tables Tables 1-3.

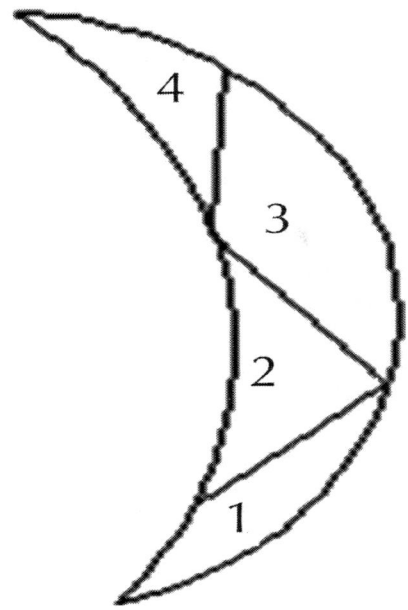

Figure 2: Typical spherical shell mesh.

Table 1: Corresponding equivalent maximum Von-Mises stress, thickness and factor of safety internal pressure of 10.0 MPa

	Thickness $_{(m)}$	Developed Stress × 10⁹ (N/m²)	V o n - M i s e s Stress × 10⁹ (N/m²)	Factor of Safety Based on Von-Mises
Element 1	0.0039	0.1286	0.1818	1.32
Element 2	0.0039	0.1274	0.1802	1.33
Element 3	0.0039	0.1235	0.1747	1.37
Element 4	0.0039	0.1278	0.1807	1.33

Table 2: Corresponding equivalent maximum Von-Mises stress, thickness and factor of safety internal pressure of 20.0 MPa

	Thickness (m)	Developed Stress × 10^9 (N/m²)	Von-Mises Stress × 10^9 (N/m²)	Factor of Safety Based on Von-Mises
Element 1	0.0079	0.1239	0.1752	1.36
Element 2	0.0079	0.1316	0.1861	1.29
Element 3	0.0079	0.1246	0.1762	1.36
Element 4	0.0079	0.1278	0.1808	1.33

Table 3: Corresponding equivalent maximum Von-Mises stress, thickness and factor of safety internal pressure of 30.0 MPa

	Thickness (m)	Developed Stress × 10^9 (N/m²)	Von-Mises Stress × 10^9 (N/m²)	Factor of Safety Based on Von-Mises
Element 1	0.0059	0.1281	0.1811	1.33
Element 2	0.0059	0.1317	0.1862	1.29
Element 3	0.0059	0.1304	0.1844	1.30
Element 4	0.0059	0.1277	0.1807	1.33

The thicknesses calculated are the minimum thicknesses required by the spherical shell to withstand the internal pressure. It is required of the design engineer to have the overall idea of all the possible loads/ forces to act on the spherical pressure vessels. And also, the overall effects of the combination of various loads/forces.

In some cases, it is advisable to add corrosion allowance to the minimum calculated thickness. The value of corrosion allowance depends on many factors. Some of the factors are:

- The corrosion rate of the stored gas/fluid on the shell material.
- Design life of the spherical pressure vessel.

- External corrosion agents at the site.
- Corrosion control and prevention methods adopted.

Having put all these into consideration by the design engineer, minimum shell thickness due to various loads/ forces can then be determined. It is of notes that if this shell thickness is lower than the minimum shell thickness recommended by ASME code, minimum shell thickness recommended by ASME code has to be used.

REFERENCES

1.   P. E. Grafton and D. R. Strome, "Analysis of Axis-Symmetric Shells by the Direct Stiffness Method," AIAA Journal, Vol. 1, No. 10, 1963, pp. 2342-2347. doi:10.2514/3.2064

2.   R. E. Jones and D. R. Strome, "Direct Stiffness Method Analysis of Shells of Revolution Utilizing Curved Elements," AIAA Journal, Vol. 4, No. 9, 1966, pp. 1519- 1525.doi:10.2514/3.3729

3.   C. Brebbia and J. J. Connor, "Stiffness Matrix for Shallow Rectangular Shell Element," Journal of the Engineering Mechanics Division, Vol. 93, No. 5, 1967, pp. 43- 65.

4.   G. Cantin and R. W. Clough, "A Curved Cylindrical Shell Finite Element," AIAA Journal, Vol. 6, No. 6, 1968, pp. 1057-1062. doi:10.2514/3.4673

5.   A. B. Sabir and A. C. Lock, "A Curved Cylindrical Shell Finite Element," International Journal Mechanical Science, Vol. 14, No. 2, 1972, pp. 125-135. doi:10.1016/0020-7403(72)90093-8

6.   Christchurch Convention Centre, Christchurch, 2006. http://www.contech.co.nz/uploaded/Post-tensioned%20LNG%20Storage%20Tanks.pdf

7.   Q. S. Chen, J. Wegrezyn and V. Prasad, "Analysis of Temperature and Pressure Changes in Liquefied Natural Gas (LNG) Cryogenic Tanks," Cryogenics, Vol. 44, No. 10, 2004, pp. 701-709. doi:10.1016/j.cryogenics.2004.03.020

8.   S. J. Jeon, B.-M. Jin and Y.-J. Kim, "Consistent Thermal Analysis Procedure of LNG Storage Tank," Structural Engineering and Mechanics, Vol. 25, No. 4, 2007, pp. 445-466.

9.   B. T. Oh, S. H. Hong, Y. M. Yang, I. S. Yoon and Y. K. Kim, "The Development of KOGAS Membrane for LNG Storage Tank," Proceedings of the 13th International Offshore and Polar Engineering Conference, Honolulu, May 25-30 2003, pp. 441-446.

10.   M. Graczyk, T. Moan and O. Rognebakke, "Probabilistic Analysis of Characteristic Pressure for LNG Tank," Journal of Offshore Mechanics and Arctic Engineering, Vol. 128, No. 2, 2005, pp. 128-133.

11.   S. R. Gorlar and O. Haddad, "Finite Element Heat Transfer and Structural Analysis of a Cone-Cylinder Pressure Vessel," International Journal of Applied Mechanics and Engineering, Vol. 12, 2007, pp. 951-963.

12.   O. Adeyefa and O. Oluwole, "Finite Element Analysis of Von-Mises Stress Distribution in a Spherical Shell of Liquefied Natural Gas (LNG) Pressure Vessels," Engineering, Vol. 3, 2011, pp. 1012-1017. doi:10.4236/eng.2011.310125

13.   O. C. Zienkiewicz and R. L. Taylor, "The Finite Element Method," 5th Edition, Vol. 2, Solid Mechanics, Butterworth-Heinemann, Oxford, 2000.

14.   E. Reissner, "On Some Problems in Shell Theory, Proceedings of the 1st Symposium on Naval Structural Mechanics," Stanford University, Pergaman Press Inc., New York, 1958.

15.   W. H. Bowes and L. T. Russell, "Stress Analysis by the Finite Element Method for Practicing Engineers," Lexington Books, Lexington, 1975.

Chapter 4

Effect of Pretreatment of Sulfide Refractory Concentrate with Sodium Hypochlorite, Followed by Extraction of Gold by Pressure Cyanidation, on Gold Removal

Alejandro Valenzuela[1,2], Jesús L. Valenzuela[1], and José R. Parga[2]

[1]Department of Chemical Engineering and Metallurgy, University of Sonora, Hermosillo, México

[2]Department of Metallurgy and Materials Science, Institute Technological of Saltillo, Saltillo, México

ABSTRACT

The majority of the refractory gold and silver occurs in occlusion in sulphides, then precious metal dissolution is possible by first oxidizing auriferous sulfide concentrate using sodium hypochlorite-sodium-hydroxide solution followed of pressure cyanidation of the oxidized concentrate, for the extraction of precious metals. This process was conducted and evaluated under cyanide and oxygen pressure. This versatile approach offers many advantages, including low temperatures, low pressure and less costly materials of construction than conventional pressure oxidation. In this study, the effect of oxygen pressure, concentration of sodium hypochlorite, temperature, and initial pH, in precious metals recovery and As removal were evaluated using a 2^4 factorial design. Characterization of the ores showed that pyrite and arsenopyrite were the main minerals present on the concentrate. The best results for gold extraction were obtained with oxygen pressure of 80 psi, 10% (w/w) sodium hypochlorite, temperature of 80°C, at pH = 13, and a constant stirring speed of 600 rpm. These conditions allowed an approximated 60% of gold and 90% of silver extractions in 1 hr.

INTRODUCTION

Background

Many of the gold deposits contain finely disseminated gold in iron sulfide minerals such as pyrite and arsenopyrite. Such deposits are called refractory gold ores due to the encapsulation of fine particles of gold in the ore host. A clear boundary between refractory and non-refractory ore in terms of gold recovery is not well defined, but several authors have pointed out that less than 60% gold recovery by direct cyanide leaching, after fine grinding of mineral indicates a refractory mineral [1]. The refractory ore must be destroyed or attacked by chemical means with the use of oxidative processes, such as oxidation by roasting [2], pressure oxidation [3], bio-oxidation [4], and ultrafine grinding [5]. Then, a suitable pretreatment process is often required to overcome the refractoriness and render the gold and silver accessible to the lixiviante action of cyanide and oxygen [1].

Oxidative leaching of metal sulphide minerals is commonly considered as an electrochemical reaction between sulfur and oxygen ion. The sulfide sulfur is oxidized, either in elemental sulfur, sulfite, or sulfate ions. Oxygen undergoes a reduction reaction with formation of water. Studies show two competing reactions, the reaction of anodic oxidation of pyrite, one of them yielding sulfate and sulfuric acid 1) and the other produces sulfate and elemental sulfur 2). Oxygen can also directly oxidize the pyrite producing Fe^{3+} 3) and thiosulfate decomposition followed by oxidation of thiosulfate and sulfite ions to sulfate [6].

$$FeS_2 + 7/2O_2 + H_2O \rightarrow FeSO_4 + H_2SO_4 \tag{1}$$

$$FeS_2 + 2O_2 \rightarrow FeSO_4 + S^0 \tag{2}$$

$$4FeS_2 + 7O_2 + 4H^+ \rightarrow 4Fe^{3+} + 4S_2O_3^{2-} + 2H_2O \tag{3}$$

The proposed reactions of arsenopyrite are quite similar (Equations (4) and (5)), [7].

$$4FeAsS + 13O_2 + 6H_2O \rightarrow$$
$$4H_3AsO_4 + 4FeSO_4 \tag{4}$$

$$4FeAsS + 7O_2 + 2H_2O + 4H_2SO_4 \rightarrow$$
$$4H_3AsO_4 + 4FeSO_4 + 4S^0 \tag{5}$$

Pressure oxidation typically operates at moderate temperature and pressure, temperatures of 60°C to 80°C and an oxygen pressure of 60 to 80 psi. Under these conditions, the sulfide sulfur is oxidized to sulfate.

Alkaline Oxidation

Arsenopyrite has an economic importance, when there is more of it than gold in the ore. Oxidation of arsenopyrite is of practical and theoretical importance in processing gold ores and concentrates. However, the chemistry of arsenopyrite in relation to its dissolution, flotation, and electrochemistry has received little attention [6]. Therefore, in alkaline oxidizing conditions, arsenopyrite can be oxidized to ferrous hydroxide, arsenate ($HAsO_4^{2-}$ or AsO_4^{2-} depending on pH), and to sulfate with iron hydroxides and arsenites as intermediates. The overall reaction for the alkaline oxidation of arsenopyrite might be described by the following equation:

$$2FeAsS + 10OH^- + 7O_2 \rightarrow$$

$$Fe_2O_3 + 2SO_4^{\ 2-} + 2AsO_4^{3-} + 5H_2O$$

(6)

Studies by Poling and Beattie [6] indicate that the oxidation of arsenopyrite, forms iron hydroxide films on the surface of the mineral at pH values above 7. Bhakta et al. [6] found that the lower oxidation potential of arsenopyrite occurs in caustic solutions. Research the electrochemical behavior of arsenopyrite in alkaline media using cyclic voltammetric at pH 8 - 12. Their work suggests that the anodic oxidation of the arsenopyrite is a two-phase dissolution mechanism. The first step is:

$$FeAsS + 6H_2O \rightarrow$$

$$Fe(OH)_3 + H_2AsO_3^- + S^0 + 7H^+ + 6e^-$$

(7)

with the formation of: a surface layer of FeOOH, elemental sulfur, and arsenite (H_2AsO_3). The second step involves the oxidation of arsenite (H_2AsO_3) to arsenate ($HAsO_4^{2-}$) and sulfate sulfur according to the following reactions:

$$H_2AsO_3^- + H_2O \rightarrow HAsO_4^{2-} + 3H^+ + 2e^-$$

(8)

$$S^0 + 4H_2O \rightarrow SO_4^{2-} + 8H^+ + 6e^-$$

(9)

At higher potentials, the oxidation products of arsenopyrite; are elemental sulfur and arsenate in accordance with the following overall equation:

$$FeAsS + 7H_2O \rightarrow Fe(OH)_3 + HAsO_4^{2-} + S^0 + 10H^+ + 8e^-$$

(10)

In the oxidation, ferric hydroxide and elemental sulfur precipitates on the surface of the particle. These compounds significantly reduce gold extraction during the subsequent recovery step by cyanidation of the neutralized residue. There are few studies on the chemical leaching of arsenopyrite in alkaline oxidizing media. The oxidation reactions, which take place in NaOH aqueous solutions, are suggested as follows:

$$3FeAsS + 9NaOH + 4O_2 \rightarrow$$
$$2Na_3AsO_4 + Na_3AsS_3 + 3Fe(OH)_3$$

(11)

Subsequent oxidation of AsS_3^{-3} produces SO_4^{-2} and AsO_4^{-3}.

Arsenopyrite oxidation rate depends on the concentrations of oxygen and alkali, temperature, surface or size of mineral particles, the pulp density of the mixture, stirring and reaction time.

Based on studies, on the published data and on experience, it can be assumed that the oxidation of pyrite and arsenopyrite in sodium hypochlorite solutions, are made in strongly alkaline conditions and high potential conditions, close to 1.2 V, a region where hypochlorite ion is stable, according to Eh-pH diagram, Figure 1" target="_self">

Figure 1(b) (H_2O-O_2-Cl_2) and in Figure 1" target="_self"> Figure 1(a), (Fe-As-S-O-H_2O) stable species are observed for iron under these conditions; therefore the oxidation of pyrite and arsenopyrite, are given according to the following reactions:

Arsenopyrite

$$FeAsS + 5NaClO + O_2 + 2NaOH \rightarrow$$
$$FeAsO_4 + Na_2SO_4 + 5NaCl + H_2O$$

(12)

$$FeAsS + 4NaClO + 1.5O_2 + 5NaOH \rightarrow$$
$$Na_3AsO_4 + Na_2SO_4 + Fe(OH)_3 + 4NaCl + H_2O$$

(13)

Pyrite

$$FeS_2 + 5.5NaClO + O_2 + 4NaOH \rightarrow$$
$$Fe(OH)_3 + 2Na_2SO_4 + 5.5NaCl + 0.5H_2O$$

(14)

Hypochlorite as Oxidizing Medium

Although the conventional direct cyanidation has been used commercially for extraction of precious metals, several studies are being conducted on alternative processes, as there are minerals that do not respond satisfactorily, classified as refractory ores, that require long leach without reaching satisfactory results. To achieve the extraction of precious metals several pretreatments are required.

For extraction of gold from refractory ores or concentrate, in which gold is surrounded or covered by iron sulfide minerals, mainly arsenopyrite and pyrite, mineral oxidative pretreatment is essential to oxidize sulfides and to expose gold to the leaching solution. Pyrometallurgical oxidation by roasting is under environmental restrictions [1]. Therefore, hydrometallurgical methods are preferred as they have better potential. These methods include pressure oxidation [7,8], biological oxidation [9- 11], and nitric acid oxidation [12].

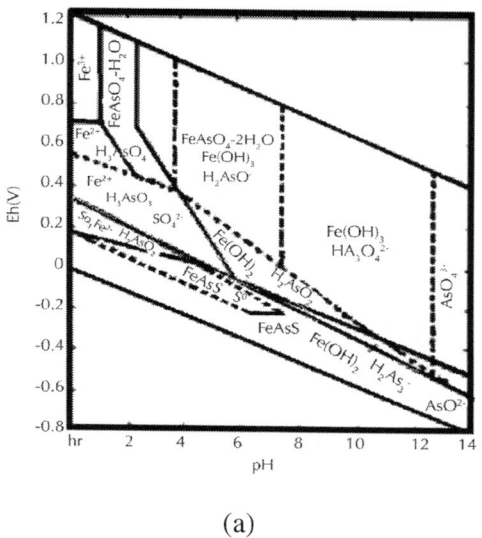

(a)

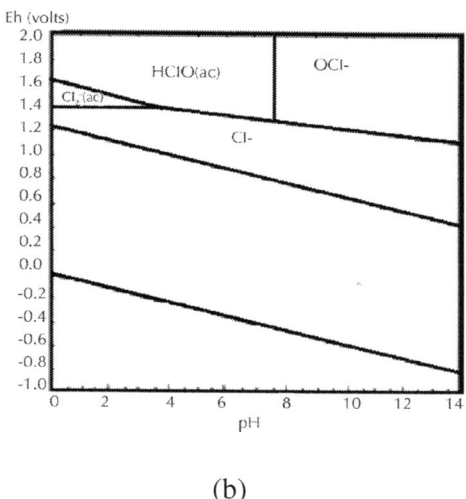

(b)

Figure 1: (a) Eh-pH Diagram of Fe-As-S-O-H$_2$O system [1]; (b) Eh-pH Diagram for H$_2$O-O$_2$-Cl$_2$ system [6].

Chlorine gas in aqueous solutions, depending on the pH, can form three oxidizing species: aqueous chlorine ($Cl_{2(aq)}$), hypochlorous acid (HOCl) and hypochlorite ions (OCl-). Concentration of these species is determined by the following equations and equilibrium constants:

$$OCl^- + H^+ \rightarrow HClO \quad K = 10^{7.5}$$

(15)

$$Cl_{2(aq)} + H_2O \rightarrow HClO + H^+ + Cl^-$$

$$K = 4.1 \times 10^{-4}$$

(16)

Those three species can be generated by adding sodium hypochlorite (NaOCl) or calcium hypochlorite ($Ca(OCl)_2$) to an aqueous solution. Calcium hypochlorite is more stable than sodium hypochlorite and contains a higher concentration of chlorine.

These salts ionize in water, under acidic conditions (pH < 7.5). Hypochlorite ion becomes hypochlorous acid in very acidic conditions (pH < 3.5), and in the presence of chloride ions, aqueous chlorine is formed. Therefore, pH should be maintained in the range of stability of OCl-, Figure 1" target="_self"> Figure 1(b). In hypochlorite solutions, in the pH range in which hypochlorite ion (OCl-) prevails, all sulfides commonly associated with gold readily oxidize. Pyrite is the most stable of metal sulfides, this means that if in aqueous solution, pyrite is oxidized; other sulfides will surely be oxidized. Pourbaix diagram of Fe-S, shows that pyrite decomposes into iron hydroxide and sulfate in the range of the hypochlorite ion. Oxidation of sulfides in hypochlorite solutions in alkaline medium has an advantage, which is sulphate formation instead of elemental sulfur that can passivate the surface of the mineral particles [13].

MATERIALS AND METHODS

Materials

The ore used was provided by Williams Mining, it was a complex of sulphides (pyrite and arsenopyrite) from a flotation bulk concentrate.

Reagents

Among others, the following regents were used; sodium hypochlorite (NaOCl), sodium hydroxide (NaOH) as pH adjuster, oxygen as the oxidant in the pressure reactor and sodium cyanide (NaCN).

Equipment

The following equipment was used: a one liter stainless steel reactor (Parr, series 4520) to carry out Alkaline oxidation pretreatment and pressurized cyanidation; a Perkin Elmer Analyst 400 atomic absorption spectrometer to determine concentration of gold, silver and arsenic, for the fire analysis, muffles, hot plates, and a variety of glassware were used, an X-ray Diffraction Microscope (XRD) and a JEOL 5410LV, Tokyo, Japan Scanning Electron Microscope (SEM) for characterization of samples.

EXPERIMENTAL METHODOLOGY

The first step was characterization of the ore and oxidized solid product.

Next, a diagnostic test of stirred conventional cyanidation was performed, under the following conditions: 100 g of sample, 0.10% NaCN, pulp to 15% (w/w), pH = 11 (adjusted with NaOH) and 72 hours treatment, and with mechanical stirring at 750 rpm, resulting in a 6% gold and 30% silver recovery.

In order to evaluate statistically the extraction of gold, a 2^4 factorial design was used. Pretreatments were run randomly and according to the conditions indicated in Table 1. The factors and levels considered were: A, concentration of sodium hypochlorite (high 10% and low 5% w/w); B, oxygen pressure (high 80, low 60 psi); C, temperature (high 80, low 60°C) and D, pH (high 13 and low 12).

For the pressurized oxidant alkaline cyanidation, all tests were performed at the following conditions: 10 g of NaCN, 60°C, 60 psi oxygen pressure, pH 11, 15% (w/w) slurry and a stirring speed of 600 rpm. Before cyanidation, the oxidized ore was washed until the complete elimination of oxidizing reagents.

RESULTS

Characterization of the Ore and Oxidized Solid Product

Table 2 shows the content of precious metals and arsenic of the concentrate, and Figure 2 shows the X-ray diffraction analysis, in which mineralogical species has been properly identified.

Table 1: Experimental design matrix

Run	A	B	C	D	Run	AB		C	D
1,17	-	-	-	-	9,25	-	-	-	+
2,18	+	-	-	-	10,26	+	-	-	+
3,19	-	+	-	-	11,27	-	+	-	+
4,20	+	+	-	-	12,28	+	+	-	+
5,21	-	-	+	-	13,29	-	-	+	+
6,22	+	-	+	-	14,30	+	-	+	+
7,23	-	+	+	-	15,31	-	+	+	+
8,24	+	+	+	-	16,32	+	+	+	+

Table 2: Chemical analysis of the concentrate

ELEMENT	AMOUNT
Au	21.01 g/t
Ag	155 g/t
As	11.20%
S	36%
Cu	0.14%
Fe	39.49%
Mn	14.20%
Pb	0.60%

Figures 3 and 4 show SEM micrographs of the ore particles, respectively, before and after the oxidizing treatment with sodium hypochlorite. Showing that the ore particles have a smooth shiny surface and oxidized ore particles show that the surface has been altered, presenting less bright, small pores, and noticing deposits of iron salts, such as sulfate and hydroxide.

Figure 4 shows the oxidizing action of sodium hypochlorite using sodium hydroxide as a pH regulator. The final solids weight was increased due to the formation of stable compounds that precipitate as sulfate and iron hydroxide.

Characterization of Oxidized Concentrate Products

Figure 5 shows the X-ray Diffractogram of the concentrate product from oxidation. It was observed that sulfur is removed as well as arsenic.

As, Au and Ag analysis of Treated Refractory Gold Concentrate

Once the experimental design was run on the 32 refractory concentrate samples, the next step was to determine the amount of gold removed. The pretreated ore was then taken to the pressure cyanidation stage, in order to analyze Au, Ag and As in the pregnant solution. The percentage of gold, silver and arsenic removal of the treated concentrate is presented in Table 3.

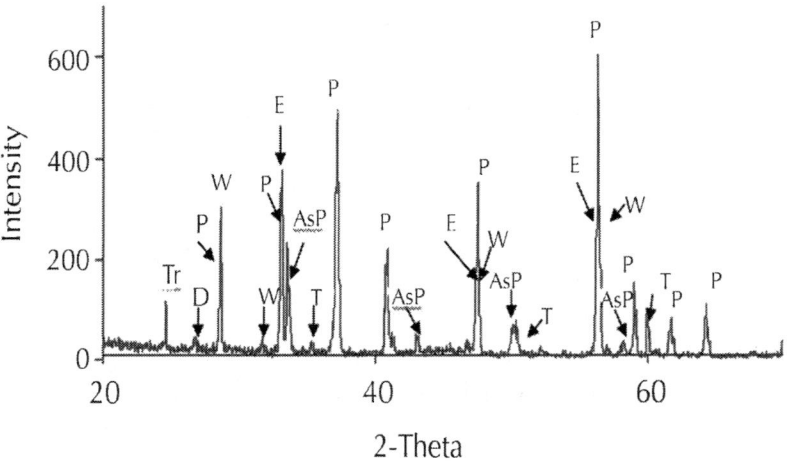

Figure 2: X-ray Diffractogram, with the mineralogical species found in the ore, AsP = Arsenopyrite (FeAsS); W = Wurtzite (ZnS); E = Sphalerite (ZnS); P = Pyrite (FeS$_2$); T = Tennantite ((Cu,Fe)$_{12}$As$_4$S$_3$); D = Domeykite-β(Cu$_3$As); Tr = Trechmannite (AgAsS$_2$).

Figure 3: Micrograph of concentrate particles.

(a)

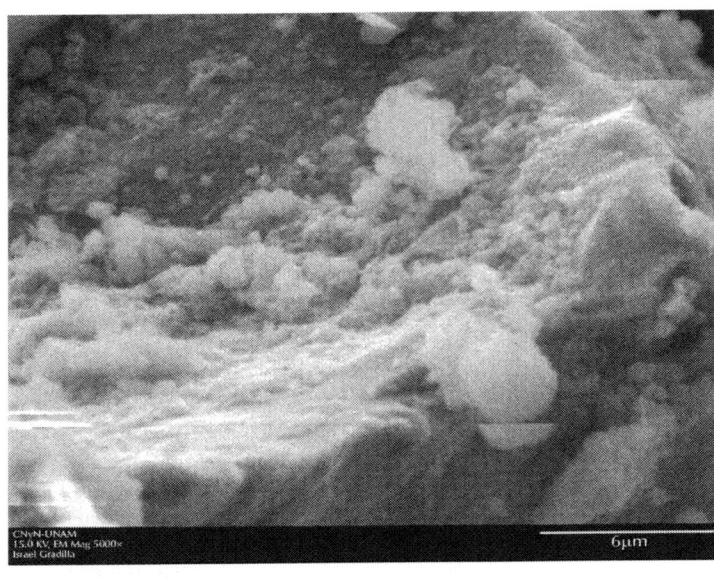

(b)

Figure 4: Micrographs of oxidized concentrate particles with NaClO.

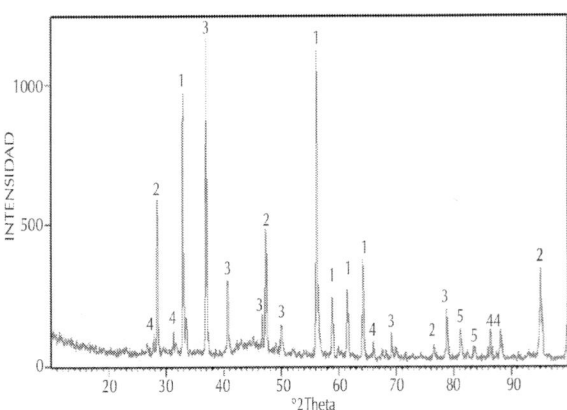

Figure 5: X-ray diffractogram, with mineralogical species found in the oxidized concentrate; 1. pyrite (FeS_2); 2. iron hydroxide ($Fe(OH)_3$); 3. arsenopyrite (FeAsS); 4. iron sulfate ($FeSO_4$); and 5. Trechmannite ($AgAsS_2$).

Graphical Analysis of Results

The statistical analysis was made using MINITAB. Main effects for gold removal, are shown graphically in Figure 6, (interactions are not shown). The biggest effect corresponds to the oxygen pressure, closely followed by NaClO concentration, then temperature, and last pH.

Formulation of the Mathematical Model

The information provided for the experimental design and analysis, allows to determine a mathematical model that can be used for prediction. It expresses the response variable—gold removal—(Y) as a function of the factors tested.

$$Y = 51.36 + 4.975A + 5.007B + 3.987C + 3.187D$$

Table 3: % of gold, silver and arsenic removal in the pretreated concentrate

Rim	As % Removal	An % Extraction	Ag % Extraction
1	13.8	32.5	83.5
2	21.7	48.7	86.5
3	18.7	46.5	83.7
4	37.8	58.7	86.9
5	20.1	39.2	87.7
6	24.9	51.5	89.5
7	30.1	58.5	85.9
8	44.3	61.9	89.9

9	30.5	40.8	88.9
10	36.7	53.7	90.6
11	37.5	52.7	89.5
12	46.1	59.3	90.9
13	36.8	48.5	87.3
14	43.3	56.1	85.6
15	44.8	59.5	89.7
16	51.0	66.2	90.6

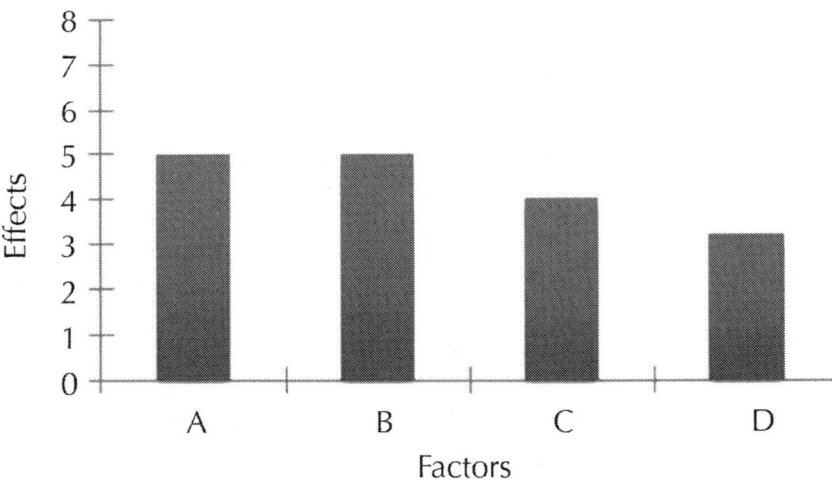

Figure 6: Graph of effect of factors on the recovery of gold, A = sodium hypochlorite concentration; B = oxygen pressure; C = temperature and D = pH.

This equation has an error (difference between the experimental extraction and calculated extraction) not greater than 3%. Therefore, the equation can be used to predict the percentage of gold extraction at any other level of the factors analyzed, within the rank tested.

CONCLUSIONS

A preliminary alkaline pretreatment of the concentrate, using hypochlorite and pure oxygen as oxidant, increases the level of extraction of gold from 6% to 60% in one hour. Preliminary oxidant alkaline processing, increases the level of extraction of gold, due to the partial oxidation of sulfur and gold exposure to the leaching solution.

Results from the experimental factorial design indicate that the four studied factors affect significantly the gold, silver and arsenic removal. The pressure of oxygen is the most important factor, followed closely for the hypochlorite concentration, next temperature and last pH. It is possible that the temperature effect is aliased, since it is related to the concentration of dissolved oxygen.

Oxidation of pyrite-arsenopyrite concentrate is improved by the increase of temperature and oxygen pressure, and these are related to the amount of dissolved oxygen. If these factors are high enough and their load size is small, the oxidation is complete within 30 to 60 minutes. The higher the temperature and oxygen pressure, the faster the oxidation.

The main advantage of the concentrated alkaline oxidation of pyrite-arsenopyrite gold is the smooth operation due to temperature and pressure, for they require less energy than acidic oxidation. Furthermore, the oxidation residues can lead to direct cyanidation, without requiring pH adjustment. The residues from cyanidation contain hydroxides and iron sulphates, which are not dangerous. Operating conditions of temperature and pressure do not require special materials for equipment.

ACKNOWLEDGEMENTS

The authors would like to thank Professors Hector Moreno and Andrew Gomes for his constructive advice during our research. This work was supported by the National Council of Science and Technology (CONACYT), University of Sonora and the Direction General of Superior Technological Education (DEGEST) from Mexico.

REFERENCES

1.   J. O. Marsden and C. I. House, "The Chemistry of Gold Extraction," 2nd Edition, S.M.E., Littleton, 2006, pp. 147-177.

2.   J. H. Coronado, M. A. Encinas, J. C. Leyva, J. L. Valenzuela, A. Valenzuela and G. T. Munive, "Tostación de un Concentrado Refractario de oro y Plata. Revista de Metalurgia," España, Vol. 48, No. 3, 2012, pp. 165-174.

3.   L. Rusanen, J. Aromaa and O. Forsen, "Pressure Oxidation of Pyrite-Arsenopyrite Refractory Gold Concentrate," Academic Journal, Physicochemical Problems of Mineral Processing, Vol. 49, No. 1, 2013, p. 101.

4.   M. A. Márquez, J. D. Ospina and A. L. Morales, "New Insights about the Bacterial Oxidation of Arsenopyrite: A Mineralogical Scope," Minerals Engineering, Vol. 3, 2012, pp. 248-254. doi:10.1016/j.mineng.2012.06.012

5.   M. Saba, M. A. R. Fereshteh and J. Moghaddam, "Diagnostic Pre-Treatment Procedure for Simultaneous Cyanide Leaching of Gold and Silver from a Refractory Gold/Silver Ore," Minerals Engineering, Vol. 24, No. 15, 2011, pp. 1703-1709. doi:10.1016/j. mineng.2011.09.013

6.   G. H. Mehdi, F. R. Schreiberb and R. Shahram, "Simultaneous Sulfide Oxidation and Gold Leaching of a Refractory Gold Concentrate by Chloride-Hypochlorite Solution," Minerals Engineering, 2012, pp. 1-3.

7.   S. A. Awe and A. Sandstrom, "Selective Leaching of Arsenic and Antimony from a Tetrahedrite Rich Complex Sulphide Concentrate Using Alkaline Sulphide Solution," Minerals Engineering, Vol. 23, 2012, pp. 1227-1236 doi:10.1016/j.mineng.2010.08.018

8.   G. C. Jones, M. Becker, P. van H. Robert and T. L. H. Susan, "The Effect of Sulfide Concentrate Mineralogy and Texture on Reactive Oxygen Species (ROS) Generation," Applied Geochemistry, Vol. 29, 2013, pp. 199-213. doi:10.1016/j.apgeochem.2012.11.015

9.   J. Jin, S. Shi, G. Liu, Q. Zhang and W. Cong, "Arsenopyrite Bioleaching by Acidithiobacillus ferrooxidans in a Rotatingdrum Reactor," Minerals Engineering, Vol. 39, 2012, pp. 19-22. doi:10.1016/j.mineng.2012.07.018

10.   J. J. K. Gordon and E. K. Asiam, "Influence of MechanoChemical Activation on. Biooxidation of Auriferous Sulphides," Hydrometallurgy, Vol. 115-116, 2012, pp. 77-83.doi:10.1016/j. hydromet.2011.12.014

11.   A. Hol, R. D. van der Weijden, G. Van Weert, P. Kondos and J. N. Cees, "Bio-Reduction of Elemental Sulfur to Increase the Gold Recovery from Enargite," Hydrometallurgy, Vol. 115-116, 2012, pp. 93-97. doi:10.1016/j.hydromet.2012.01.003

12.   G. R. Zárate, G. T. Lapidus and R. D. Morales, "Aqueous Oxidation of Galena and Pyrite with Nitric Acid at Moderate Temperatures," Hydrometallurgy, Vol. 115-116, 2012, pp. 57-63. doi:10.1016/j. hydromet.2011.12.010

13.   A. Basua and M. E. Schreiber, "Arsenic Release from Arsenopyrite Weathering: Insights from Sequential Extraction and Microscopic Studies," Journal of Hazardous Materials, Vol. 244, 2013.

5

Application of Particle Extraction Process at the Interface of Two Liquids in a Drop Column— Consideration of the Process Behavior and Kinetic Approach

Jacqueline V. Erler[1], Tom Leistner[2], and Urs A. Peuker[1]

[1]Institute of Mechanical Process Engineering and Minerals Processing, Technical University Bergakademie Freiberg, Freiberg, Germany

[2]Helmholtz-Institute Freiberg of Resource Technology, Freiberg, Germany

ABSTRACT

The focus of this research is a new type of particle extraction process for the transfer of magnetite nanoparticles from an aqueous to an immiscible organic phase, directly through the liquid-liquid phase boundary in a drop column. The particle extraction process comprises several advantages such as a minimum amount of stabilizing surfactant, no exposure of the particles to a gas atmosphere and with it the avoidance of sintering by capillary forces and a high particle concentration in the receiving phase as well. The study presents experimental results of the characterization of the process environment and the transfer behavior in a drop column. The solution of surfactant in the continuous phase has been investigated during a particle-free phase transfer experiment including the measurements of the total organic carbon (TOC) content and analysis of the size of the stabilized droplets using the laser diffraction spectroscopy. The determination of the transfer fluxes; the mass flows as well as the yield of transferred magnetite by ICP-OES measurements provide information on the impact of interaction of the elementary processes at the phase boundary. Furthermore, the transfer kinetics of the process is described and compared with calculated theoretical values resulting from a kinetic approach.

INTRODUCTION

Because of their special magnetic properties, magnetite nanoparticles have a great potential for many technological applications. Therefore, they are very interesting for a broad range of research areas, for example as magnetic fluids for low friction dynamic gasket systems as well as for the construction of vibration dampers and tweeters [1] [2]. A particular research focus represents the application of magnetite nanoparticles as advanced functional materials for surface coatings and particle composite materials [3] [4].

Possible areas of application for this can be found in the field of reaction engineering [5], in the form of magnetically separable catalyst material as well as in the biomedical sector [6] -[8] . Especially in combination with polymers, stabilized magnetite nanoparticles are required in an organic phase [9] - [12]. However, the nanoparticles are synthesized mainly in an aqueous phase and have to be placed in an organic phase through appropriate procedures. Due to their increased surface area/volume ratio, nanoparticles are susceptible to oxidation and have a great tendency to agglomerate, which may cause a loss of their special magnetic properties. As a consequence, conventional transfer strategies, based on filtration with subsequent drying and redispersion steps, can only be applied conditionally [13] . Therefore, the development of an efficient process to transfer the magnetite nanoparticles from the aqueous to an immiscible organic phase is of great interest.

Emphasis should be put on producing stable colloidal and functionalized particles continuously with a minimum use of surfactants in the liquid organic medium. To demonstrate the continuous phase transfer process via a particle extraction, we design a concept of a miniplant using a drop column as the chosen transfer device. In the literature, the bubble columns with different internals are thoroughly investigated regarding to mass transfer, flow patterns, bubble shapes and hydrodynamics, for example with applications in metallurgical, chemical, bioand petrochemical process industries [14] -[17] . In our study, a column is used in the simplest form without internals. An organic liquid as disperse phase is injected as drops through a distributor into a magnetite nanoparticle suspension, which represents the continuous aqueous phase. Already, in 1968 Lai and Fuerstenau [18] have carried out a liquid-liquid extraction of ultrafine particles to separate a mixture of alumina, water and oil, however they use for their experiments a separatory funnel to put on top of each other the two phases.

In this study, the particle extraction process is investigated and the results are presented with the drop column in a partial recirculation operation. This means that the aqueous phase is stationary and that the organic phase is in recirculation, as shown in Figure 1. This is a first step to the future operation with a closed circuit of both phases.

The process mechanism of the particle transfer in the drop column can be divided into three fundamental steps. At first, the nanoparticles agglomerate partly due to the instable conditions in the aqueous phase as already mentioned in [19]. In the drop column, after injecting the organic phase, no sedimentation takes place due to the stirring effect of the rising liquid drops. Particle drop collision occurs. At the interface, the surfactants from the organic phase interact with the hydrophilic particle surface. The surfactant molecules adsorbe and become chemically grafted [20] [21]. This leads to a hydrophobization and functionalization of the particles, which allows a phase transition of the magnetite.

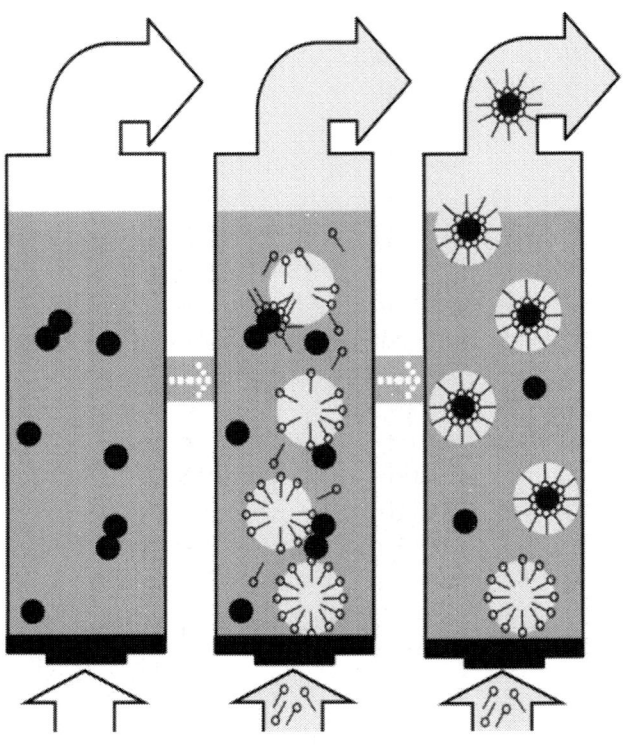

Figure 1: Scheme of the fundamental steps in the liquid-liquid particle extraction process mechanism in the drop column as transfer device.

Depending on the surfactant used, stable organic colloids are formed by an excellent deagglomeration of the particles and a physicochemical dispersion due to the strong repulsive potential of the adsorbed surfactant molecules [12].

The aims for the phase transfer process in the drop column are a high efficiency, which means a high yield of transferred magnetite in the organic phase, low transfer times or rather process times and a stable product. Furthermore, a stable process in the drop column is necessary. It means a stable drop formation and upward movement with a sufficient coalescence rate, whereby a clearing off of the column is essential for the setup chosen.

MATERIALS AND METHODS

The synthesis of the magnetite nanoparticles with a crystallite size of about 15 nm [21] by a wet-chemical coprecipitation reaction is carried out at 70°C under atmospheric conditions as described by Machunsky et al. [13]. Therefore we applied the precursors iron (II) sulphate heptahydrate and iron (III) chloride hexahydrate purchased from Carl Roth Germany as well as the precipitant ammonium hydroxide solution with an ammonia content of 26% from Sigma.

For the phase transfer experiments in the drop column the pH-conditioning of the aqueous phase is necessary; otherwise emulsion formation in the drop column occurs. The aqueous suspension is conditioned under ambient air by repeated washing with distilled water to a pH value of 4 - 5. For the experiments with oleic acid (OA) the original salt concentration of 39.4 g/l and with ricinoleic acid (RA) a quarter of the original salt concentration is utilized. Subsequently, for both surfactants used a process-pH value of 8.10 ± 0.05 is adjusted.

The research of the liquid-liquid particle extraction process in the drop column is affected by the choice of the surfactants, ricinoleic acid and oleic acid respectively. Both are unsaturated fatty acids (FA) at the 9th C-atom with a carbon chain length of 18 C-atoms. RA has an additional hydroxyl-group at the C-atom C12. The surfactants are of technical grade with 90% purity and purchased from Sigma.

The organic phase as the disperse phase is consisting of the solvent iso-octane provided by Carl Roth Germany with 99.5% purity as well as the mass fraction of surfactant x_{surf} and a specific amount of surfactant per magnetite $X_{S/M}$. The study presents the results with the parameters x_{surf} = 1.4 mass-% and $X_{S/M}$ = 0.2 g/g.

All chemicals are used as received.

The magnetite mass concentration is determined with ICP-OES analyses (inductively coupled plasma optical emission spectroscopy) of Fe with the ICP spectrometer iCAP 6300 from Thermo Fischer Scientific. For the measurements, 5 emission lines of Fe with different wave length ($Fe_{238.2\ nm}$, $Fe_{240.4\ nm}$, $Fe_{259.9\ nm}$, $Fe_{274.6\ nm}$, $Fe_{274.9\ nm}$) are used, which cover the whole mass concentration range. For this, the samples were chemically digested with concentrated hydrochloric acid delivered from Carl Roth Germany. For each emission line a triple determination was performed, resulting with a relative standard deviation of <1%.

The total organic carbon (TOC) content measured with the device from analytikjena multi N/C 2100s is determined by using the difference method. This means the total carbon (TC) content is analyzed. After degassing the inorganic carbon (IC) content is measured and subsequently the TOC content can be calculated. If the IC content is negligible the TC-method is used, in which the TC content is measured directly, and this value corresponds with the TOC content.

For the analysis of the size distribution of stabilized droplets by laser diffraction spectroscopy the spectrometer HELOS, manufactured by Sympatec, is used. The measurement range for this device is between 0.1 - 875 **µm.**

Experimental Setup

As mentioned the drop column as transfer device is used in a partial recirculation operation, that means the aqueous phase is stationary and the organic phase is in recirculation, as seen in Figure 2. The drop column has a length of 700 mm with an internal diameter of 25 mm. At the bottom of the column the distributor (sparger) as dispersing system is a single metal capillary with an inside diameter of 3.2 mm centrally mounted in a perforated plate. The mixing of the organic phase in the receiver tank is ensured by a mechanical agitator.

Limited by the dimensions of the transfer device a volume of 300 ml of the magnetite suspension, which corresponds to a mass of 6 g magnetite nanoparticles, is filled into the column as continuous phase. The organic phase, which acts as disperse phase is pumped from the receiver tank through the metal capillary into the drop column by a supply system consisting of a peristaltic pump (Ismatec Reglo Analog) and a Tygon-tube with an inside diameter of 3.2 mm. Directly above the opening of the metal capillary, whereby dead zones can be formed aside, the organic phase immediately disintegrates in differently sized drops. These rise through the column in the aqueous phase, due to their lower density. The drops transport the extracted magnetite and thus the nanoparticles are concentrated within the coalesced organic phase. The presence of the solvent drops within the column leads to an expansion of the liquid level, the so-called hold-up.

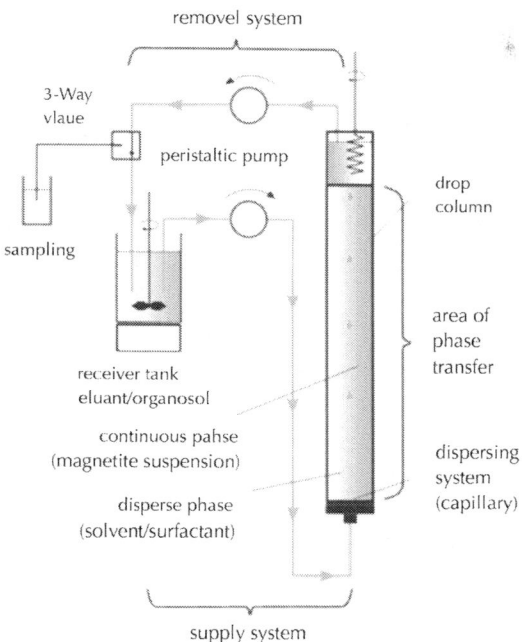

Figure 2: Experimental setup of the miniplant in a partial recirculation operation: aqueous phase stationary and organic phase in recirculation.

The organic phase containing the extracted magnetite particles is collected at the top of the liquid and pumped back to the receiver tank through the removal system. Due to the fact that surfactants are used, the rising drops can form a so called dispersion ribbon at the surface of the continuous phase. This is a layer consisting of droplets and only partial hydrophobized particles, where the coalescence is hindered [22]. The removal system comprises a peristaltic pump and Tygon-tubes with the same properties as in the supply system as well as an automated sampling system with a pneumatic 3-way valve, manufactured by Swagelok, which is controllable via measurement software.

Due to the hydrodynamic dead time the first measuring point can be taken after 85 seconds, because then organosol is existent in the removal system. The average resistence time in the column is about 51 seconds and the volume flow rate is 29 ml/min. The sample volume amount is 2.9 ml and is added at the same time manually as pure organic phase in the receiver tank, hereby keeping the volume of the disperse phase constant during the recirculation. However, after each removed sample volume the amount and mass concentration of magnetite nanoparticles in the system is reduced, which is included in the calculation of $\beta_{magn,rt}(t)$ based on the validation of the system.

For the evaluation of the corrected mass concentration of transferred magnetite nanoparticles in the receiver tank $\beta_{magn,rt}(t)$ a verification of the system is applied taking into consideration the input and output to the receiver tank as well as the residence time distribution in Equation (1):

$$\beta_{magn,rt}(t) = \beta_{magn,rt,0} \times e^{-\frac{\dot{V}_{dtsp}}{V_{rt}}t} + \int_{t'=0}^{t'=t} \frac{\dot{V}_{disp}}{V_{rt}} \beta_{magn,in}(t) \times e^{-\frac{\dot{V}_{dtsp}}{V_{rt}}(t-t')} dt \tag{1}$$

Where $\beta_{magn,rt,0}$ the initial mass concentration of the system is, V_{rt} is the volume in the receiver tank, V_{disp} is the volume flow of the disperse phase and t' is the auxiliary variable, which describes the time of the entry.

By linear interpolation and numeric integration, the Equation (1) can be solved. This results in the following equation:

$$\beta_{magn,rt}(t) = \left(\beta_{magn,rt,0} + \frac{a_i \times V_{rt}}{\dot{V}_{disp}} - b_i \right) \times e^{-\frac{V_{rt}}{\dot{V}_{disp}}t} + a_i t + b_i - \frac{a_i \times V_{rt}}{\dot{V}_{disp}} \tag{2}$$

Where a_i as well as b_i are formed by linear interpolation between the removed samples.

To characterize the process environment we have investigated the procedures in the drop column by the method of a particle-free operation, whereby the changes in the column can be observed. At defined time points the pumps were switched off and the sampling is affected from the continuous phase. In dependence from surfactants used the different mechanisms of drop formation of the disperse phase in the column is demonstrated in Figure 3. With ricinoleic acid as surfactant (Figure 3(a)) the disperse phase disintegrates in a multitude of very small droplets and turbidity appears in the column at once, which spreads in the whole column within a few minutes. In contrast, with oleic acid (Figure 3(b)) as surfactant a single periodic drop formation [23] occurs resulting in larger and significantly less drops without turbidity.

RESULTS AND DISCUSSION

As mentioned above the both surfactants used have a different process behavior with varying consequences to the particle extraction process. The turbidity in the column with ricinoleic acid as surfactant does not influence the process stability of the phase transfer, because the disperse phase can be performed in recirculation without hindrance. The reason for the formation of small emulsion droplets is due to the mass transport of ricinoleic acid through the phase boundary. This is possible, because ricinoleic acid is soluble in the organic as well as in the salts and ammonia containing aqueous phase owing to the additional hydroxyl group. This effect of the spontaneous formation of an emulsion has been described relating to a pH value dependency of Stackelberg, 1949 [24] . In our case an imbalance prevails in the column at the interface of the rising drops, since the ricinoleic acid is dissolved in the organic phase, only at beginning of the process. The system aspires an equalization of concentration by the transport of ricinoleic acid through the phase boundary into the aqueous phase. The resulting convection currents cause deformations and lead to the formation of very small droplets. Figure 4 outlines these stabilized iso-octane droplets with a median value of 5 - 10 **μm** measured by laser diffraction spectroscopy. They can only be formed if ammonium ricinoleates (carboxylates) exist at the interface by dissolving of ricinoleic acid in the aqueous phase, which definitely reduces the interfacial tension [25]. Furthermore, the smallest median value of the stabilized iso-octane droplets is reached after 20 minutes of defined time points. Due to the low buoyant force of these droplets, the turbidity is appearing in the column.

To prove the solution of surfactants from disperse into continuous phase the TOC content in the aqueous phase is determined and plotted as a function of the time, as shown in Figure 5. By additional centrifugation of the aqueous phase, a potential carbon signal of emulsified solvent can be excluded (light gray columns as short-time trial and dark gray columns as long-time trial). By using ricinoleic acid (Figure 5, top) as surfactant, a significant increase of the TOC content can be observed. After 5 minutes process time the aqueous phase exhibits an amount of dissolved ricinoleic acid in the range of 1.6 - 2.4 g/l independent of the time limit of experimental procedure. Whereas oleic acid (Figure 5, bottom) is used, the poor ability of dissolving within the aqueous phase is expressed considerably.

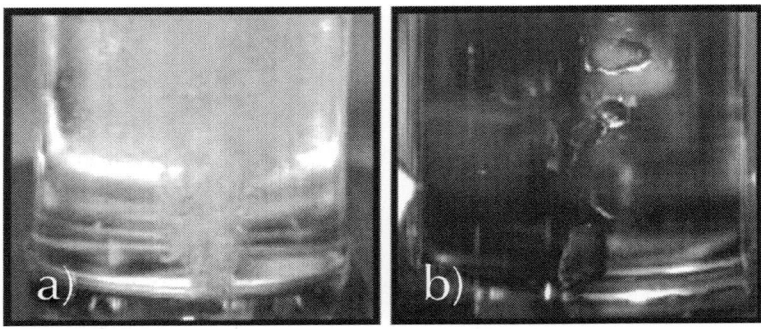

Figure 3: Drop formation of the disperse phase in the column directly above the distributor with ricinoleic acid (a) and oleic acid (b) as surfactant at beginning of the process.

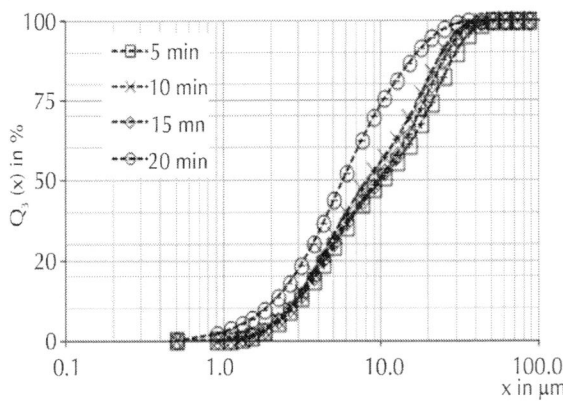

Figure 4: Size distribution of the stabilized iso-octane droplets in the aqueous phase with ricinoleic acid as surfactant using laser diffraction spectroscopy.

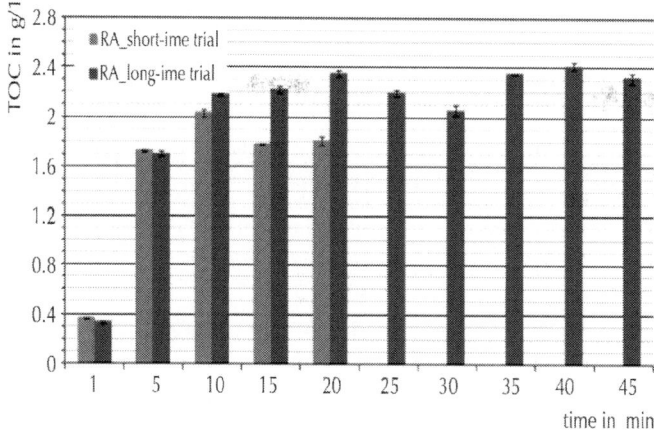

(a)

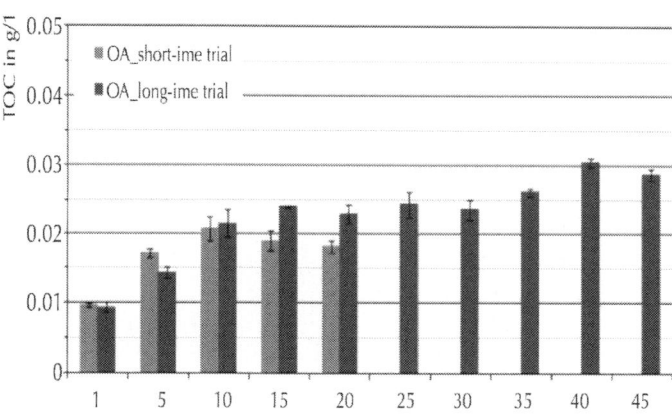

(b)

Figure 5: Total organic carbon (TOC) content at defined time points of sampling in the aqueous phase by centrifugation of the partly emulsified aqueous phase (light gray columns as short-time trials and dark gray columns as long-time trials) with ricinoleic acid (top) and oleic acid (bottom) as a surfactant.

So the TOC values of oleic acid actually dissolved in the aqueous phase are nearly in the range of 0.01-0.03 g/l. Comparing the average values of the mentioned ranges (2 g/l for ricinoleic acid and 0.02 g/l for oleic acid), it can be recognized that in contrast to oleic acid approximately the 100-fold amount of ricinoleic acid is dissolved in the aqueous phase irrespective of time.

Furthermore, experiments are performed with a pH-indicator bromothymol blue in the continuous phase, as demonstrated in Figure 6(a), which has a transition point at pH 7.6 (color change from blue to green). For either surfactant an initial process-pH value of 8.1 is adjusted, similar to the conditioned aqueous phase for the phase transfer of magnetite. Only with ricinoleic acid in the disperse phase a change from blue to green is observed in the continuous phase (Figure 6(b)). This indicates that the pH value has decreased below 7.6. The reduction in the concentration of ammonium ions is initiated by the formation of ammonium ricinoleates at the liquid-liquid interface due to the dissolving of ricinoleic acid in the aqueous phase. After the recirculation of the disperse phase with oleic acid as surfactant the color of the continuous phase has changed to light blue (Figure 6(c)), because significantly less of oleic acid is dissolved in the aqueous phase. As a result the formation of ammonium oleates at the interface is lower, which only leads to a slight change of pH in the aqueous phase.

Figure 6: Characterization of the pH value changes in the continuous phase using the indicator bromothymol blue (a) after recirculation of the disperse

phase with ricinoleic acid (b) and oleic acid (c) as a surfactant.

This finding implies that more than one process is relevant at the phase boundary.

For the characterization of the transfer behavior the yield of transferred magnetite particles $\phi_{magn,trans}$ is plotted against the process time in Figure 7. This parameter is determined, from the time dependent corrected magnetite mass concentration in the receiver tank $\beta_{magn,rt}(t)$ based on the validation of the system in Equation (1) relating to the theoretical applied magnetite mass concentration for the phase transfer $\beta_{magn,pt,theor}$ in the following equation:

$$\phi_{magn,trans} = \frac{\beta_{magn,rt}(t)}{\beta_{magn,PT,theor}}$$

(3)

Based on the short-time trial with the process time limit of about 10 minutes it can be observed excellently, that the phase transfer with ricinoleic acid as a surfactant takes place rapidly. The increase of the yield of transferred magnetite is nearly twice the amount compared to oleic acid. This is consistent with the significant increase in TOC content, because by dissolving of ricinoleic acid in the aqueous phase the drops disintegrate more easily and thus an additional phase boundary can be formed for the phase transfer process. Due to the lower solution of oleic acid in the aqueous phase the available phase interface is smaller, which leads to longer process times for the phase transfer of magnetite. This also can be observed during the long-time trials with both surfactants. With ricinoleic acid it can be demonstrated, that already after 15 minutes the yield of transferred magnetite amounts to approximately 90% and it is kept constant about over time, up to 165 minutes. However, with oleic acid the transfer is completed after nearly 270 minutes and the maximum with a yield of transferred magnetite of about 86%. Furthermore, the clearing off of the column with ricinoleic acid as surfactant is obtained after about 6 minutes, whereas with oleic acid it can be achieved only after 160 minutes.

For the characterization of transfer kinetics, the mass flow of transferred magnetite particles $m_{magn,trans}$ is plotted against the process time in Figure 8. This parameter is determined, by using the constant volume flow of the disperse phase V_{disp} and the difference in mass concentration of transferred magnetite between output and input of the disperse phase $\Delta\beta_{magn,out,in}(t)$ in Equation (4).

$$\dot{m}_{magn.trans} = \dot{V}_{disp} \times \Delta\beta_{magn,out,in}\left(t\right)$$
(4)

The plot of the mass flow with ricinoleic acid as a surfactant is conspicuously distinguished by the increase from the first measuring point at 85 seconds till the maximum of the mass flow at a process time between 4 - 5 minutes. During this initial phase the ricinoleic acid concentration in the aqueous phase still rises, which limits the mass flux of magnetite. Subsequently, the mass flow is described with a kinetic of 1st order as it is also identified with oleic acid as surfactant, what needs to be proved.

In the present case the product of functionalized magnetite particles (B) are formed from the precursor of pure magnetite nanoparticles (A) in Equation (5).

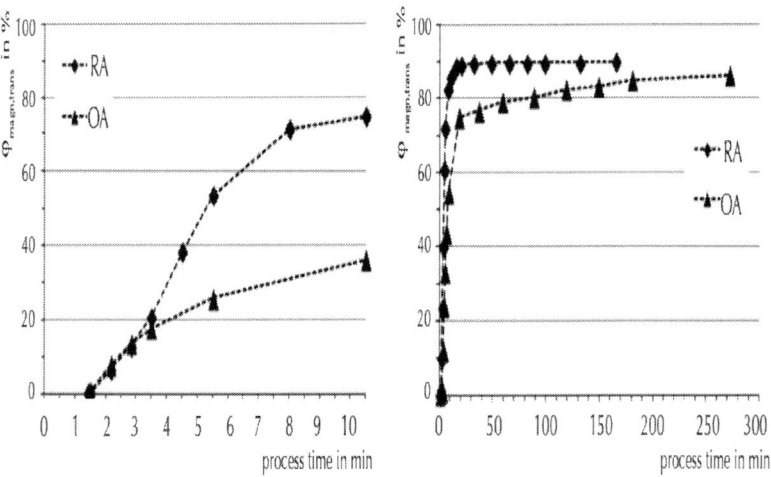

Figure 7: Yield of transferred magnetite in dependency of the process time for ricinoleic acid and oleic acid as a surfactant during short-time and long-time trials.

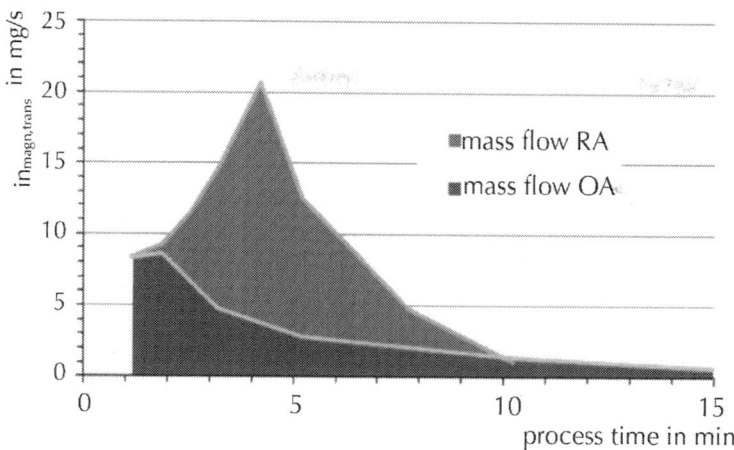

Figure 8: Demonstration of the time dependent mass flow of transferred magnetite particles dependent on the surfactant used.

$$Fe_3O_{4\,pure,aqueous}\,(A) \rightarrow FA@Fe_3O_{4\,organic}\,(B)$$

(5)

It is supposed that the already transferred magnetite concentration in the organic phase does not influence the transfer process. With this assumption the drop column can be seen as conventional tank reactor [26], in which the concentration of magnetite is reduced by the transfer reaction.

Thereby, the differential temporary law of the reaction can be established in Equation (6). This indicates, that the temporal change of mass concentration $d\beta_A/dt$ is proportional to the currently existing mass concentration β_A.

$$\frac{d\beta_A}{dt} = -k\beta_A$$

(6)

Where k is the rate constant of reaction (1/s) of the 1st order and the product $k\beta_A$ is the rate of consumption of A. Furthermore, the half-life $t_{1/2}$ for the reaction of both surfactants can be determined from the experimental data. The half-life is the time, at which half of the precursor is dissipated. By means of Figure 9 it can be distinguished, that this corresponds to the point of intersection, between the normed mass concentration of the precursor β'_A and the product β'_B. Due to the comparability a dimensionless representation of the mass concentration is chosen by the scaling in Equation (7).

$$\beta' = \frac{\beta_{magn,ICP}}{\beta_{max}}$$

(7)

The relative change in the mass concentration $d\beta_A / \beta_A$ in Equation (8) is proportional to the temporal change dt. It can be calculated from the experimental data by the ICP-OES measurements and is demonstrated in Figure 10 (plots using filled symbols).

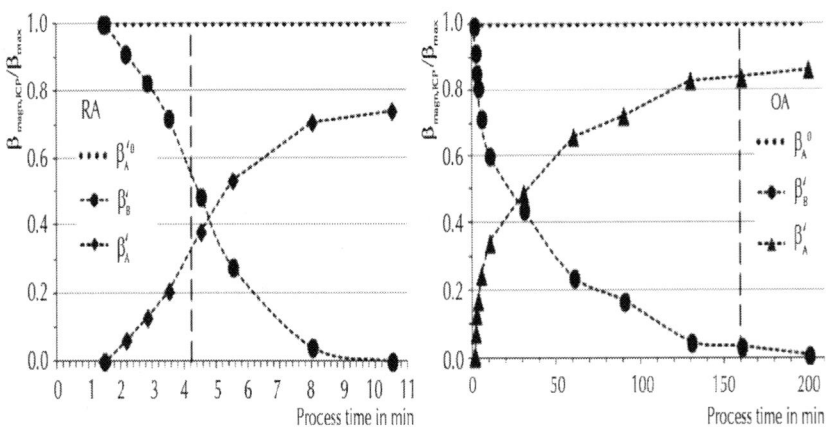

Figure 9: Dependency of the half-life of surfactant used presented by the normed mass concentration plots—with ricinoleic acid (left) the half-life is 4.78 minutes, whereas the half-life with oleic acid (right) is 27.2 minutes. The dashed line indicates the clearing off of the column.

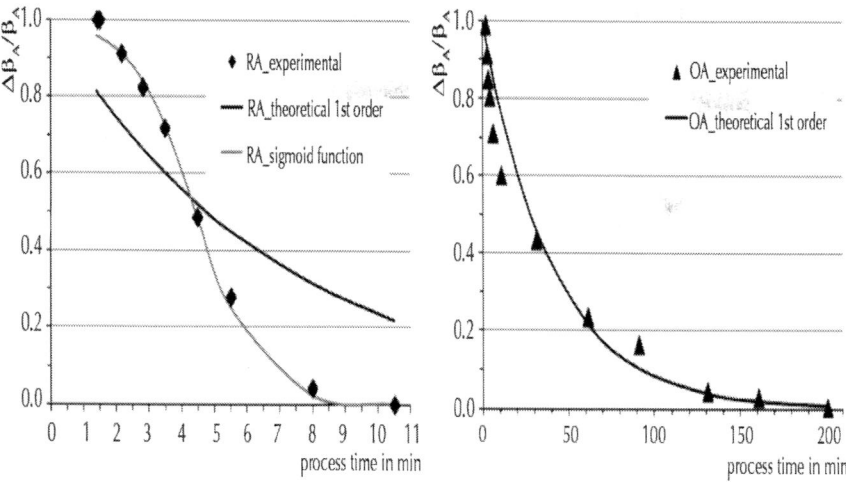

Figure 10: Demonstration of the experimental as well as theoretical calculated relative change in the mass concentration in dependency of the process time with ricinoleic acid (left) and oleic acid (right) as a surfactant.

$$\frac{d\beta_A}{\beta_A} = -kdt$$

(8)

By integration from t=0 till end time t as well as from initial mass concentration β_A^o till $\beta_A(t)$ and additional by the use of logarithmic, the integral temporary law is achieved in the following equation:

$$\beta_A = \beta_A^0 \times e^{-kt}$$

(9)

Resulting in the determination of the theoretical mass concentration β_A and using the Equation (10) for the calculation of k for both surfactants derived from Equation (9)

$$k = \frac{\ln 2}{t_{1/2}}$$

(10)

Hence, the theoretical relative change in the mass concentration can be calculated and compared to the experimental calculated plots (Figure 10).

Based on the comparison in Figure 10 it can be recognized, that the plots between the experimental and theoretical calculated relative change in mass concentration agree only for oleic acid. Therefore, the particle extraction process in partial recirculation operation of the transfer device with oleic acid as a surfactant can be approximately described as a reaction of 1st order. This indicates that the dissolution of oleic acid from the organic in the aqueous phase does not influence the process kinetics. On the contrary, the assumption of a 1st order reaction cannot be used to describe the process response if ricinoleic acid is applied as a surfactant. The course of the experimental relative change in the mass concentration resembles an s-shaped curve. Therefore to characterize the transfer kinetics we can use a sigmoidor a logistic function, respectively, which is applied for approximating saturation processes and also dose-response-systems in Pharmacology [27]. Thus, the particular process behavior for ricinoleic acid can be described using the following equation:

$$\frac{\Delta \beta_A}{\beta_A} = f(t) = A_1 + \frac{A_2 - A_1}{1 + 10^{(M-t)*p}}$$

(11)

Whereby A_1 and A_2 the theoretical lower and upper thresholds of the curve Due to the fact that possible values for $\Delta \beta_A / \beta_A$ lie in the range {0...1}, A_1 and A_2 can be defined with 0 and 1, respectively. Furthermore, Equation (11) has a negative first derivation and, more important, one point of inflection, which demonstrates the local maximum of the mass flow at a process time between 4 - 5 minutes (Figure 8). The point of inflection is represented in Equation (11) by the parameter M, which is estimated through linear interpolation. The parameter P describes the slope of the curve and is numerically calculated by iteration. Thus, the function obtained is illustrated by the grey line in Figure 10, which accurately fits to the experimental data. This demonstrates that the additional mass transport due to dissolving of ricinoleic acid from the organic into the aqueous phase has an important influence on the kinetics of the particle extraction process within the column. The surfactant mass transfer can be assumed as the dose for the process, which predominates in the first minutes of the phase transfer when equilibrium of ricinoleic acid concentration in either phase is not reached. Consequently, the response is represented by the functionalization and subsequently the phase transfer of magnetite nanoparticles. Thus, it can be finally stated out that the particle extraction process is controlled by process imbalance.

CONCLUSIONS

In this work, the particle extraction process with a drop column as transfer device in partial recirculation operation has been investigated. The material parameter of the suspension are chosen to be constant with surfactant concentration $x_{surf} = 1.4$ mass-% and the optimal specific amount of surfactant per magnetite $X_{S/M} = 0.2$ g/g. For these parameters, secondary effects like water inclusion and emulsion formation can be excluded, because a stable process behavior with a complete clearing off the column as well as a sufficient coalescence rate can be achieved. Based on the modeling of the system, we are able to evaluate the transfer fluxes. Thereby, yields of transferred magnetite nanoparticles are obtained >80%. Furthermore, for the characterization of the process environment, we have chosen a particle-free phase transfer as an indicator. In combination with optical changes in the column as well as measurements of the total organic carbon (TOC) content and the size distribution using laser diffraction spectroscopy, we have proven that ricinoleic acid helps disintegrating iso-octane. Droplets with a median particle size of 5 - 10 **µm** are formed. This is particular difference in comparison to oleic acid and this is due to the chemical structure of the surfactants. Therefore, it is possible that additional phase interfaces are formed for the phase transfer process, which further determines the transfer times of particles by reduction of the interface tension. As a consequence of this, we have described the interaction of the processes and procedures at the phase boundary. Thereby, we have a mass transport of the surfactant from disperse into continuous phase on the one hand, which determines the process time for the phase transfer, and simultaneously the phase transfer of the magnetite nanoparticles on the other hand. This interconnection is reflected in the transfer kinetics of either surfactant. Finally, we have determined that the particle extraction process with oleic acid as a surfactant can be estimated as a reaction of 1st order. Thus, the influence of the surfactant mass transfer of the process kinetics is negligible. However, in the case of ricinoleic acid as surfactant, another approximation has to be used to describe the particular process behavior. This is represented by a sigmoid function in terms of a dose response curve. Therefore, the mass transport of the surfactant due to dissolving strongly influences the process kinetics.

Advanced studies in the process development of a continuous liquid-liquid phase transfer to obtain highquality organosols will be presented soon regarding to produce stable colloids and interaction between the surfactants and organic phase used.

ACKNOWLEDGEMENTS

The authors would like to thank the German Research Foundation (Deutsche Forschungsgemeinschaft DFG) for financial support by grant PE1160/6-3 and we give thanks to Andre' Rieger as well as the laboratory technicians from Institute of Thermal Process Engineering, Environmental and Natural Products Process Engineering of TU Bergakademie Freiberg for the TOC measurements.

REFERENCES

1.   Scherer, M. and Figueiredo Neto, A.M. (2005) Ferrofluids: Properties and Applications. Brazilian Journal of Physics, 35, 718-727. http://dx.doi.org/10.1590/S0103-97332005000400018

2.   Hurlebaus, S. and Gaul, L. (2006) Smart Structure Dynamics. Mechanical Systems and Signal Processing, 20, 255-281. http://dx.doi.org/10.1016/j.ymssp.2005.08.025

3.   Dallas, P., Georgakilas, V., Niarchos, D., Komninou, P., Kehagias, T. and Petridis, D. (2006) Synthesis, Characterization and Thermal Properties of Polymer/Magnetite Nanocomposites. Nanotechnology, 17, 2046-2053. http://dx.doi.org/10.1088/0957-4484/17/8/043

4.   Teja, A.S. and Koh, P.-Y. (2009) Synthesis, Properties and Applications of Magnetic Iron Oxide Nanoparticles. Progress in Crystal Growth and Characterization of Materials, 55, 22-45. http://dx.doi.org/10.1016/j.pcrysgrow.2008.08.003

5. Hickstein, B. and Peuker, U.A. (2009) Modular Process for the Flexible Synthesis of Magnetic Beads—Process and Product Validation. Journal of Applied Polymer Science, 112, 2366-2373. http://dx.doi.org/10.1002/app.29655

6. Banert, T. and Peuker, U.A. (2007) Synthesis of Magnetic Beads for Bio-Separation Using the Solution Method. Chemical Engineering Communications, 194, 707-719.http://dx.doi.org/10.1080/00986440600992750

7. Laurent, S., Forge, D., Port, M., Roch, A., Robic, C., Vander Elst, L. and Muller, R.N. (2008) Magnetic Iron Oxide Nanoparticles: Synthesis, Stabilization, Vectorization, Physicochemical Characterization and Biological Application. Chemical Reviews, 108, 2064-2110. http://dx.doi.org/10.1021/cr068445e

8. Mahmoudi, M., Sant, S., Wang, B., Laurent, S. and Sen, T. (2011) Superparamagnetic Iron Oxide Nanoparticles (SPIONs): Development, Surface Modification and Applications in Chemotherapy. Advanced Drug Delivery Reviews, 63, 24-46. http://dx.doi.org/10.1016/j.addr.2010.05.006

9. Banert, T. and Peuker, U.A. (2006) Preparation of Highly Filled Super-Paramagnetic PMMA-Magnetite Nano Composites Using the Solution Method. Journal of Material Science, 41, 3051-3056. http://dx.doi.org/10.1007/s10853-006-6976-y

10. Kirchberg, S., Rudolph, M., Ziegmann, G. and Peuker, U.A. (2012) Nanocomposites Based on Technical Polymers and Sterically Functionalized Soft Magnetic Magnetite Nanoparticles: Synthesis, Processing, and Characterization. Journal of Nanomaterials, 2012, Article ID: 670531. http://dx.doi.org/10.1155/2012/670531

11. Rudolph, M. and Peuker, U.A. (2011) Coagulation and Stabilization of Sterically Functionalized Magnetite Nanoparticles in an Organic Solvent with Different Technical Polymers. Journal of Colloid and Interface Science, 357, 292- 299.http://dx.doi.org/10.1016/j.jcis.2011.02.043

12. Rudolph, M. and Peuker, U.A. (2012) Phase Transfer of Agglomerated Nanoparticles—Deagglomeration by Adsorbing Grafted Molecules and Colloidal Stability in Polymer Solutions. Journal of Nanoparticle Research, 14, 990. http://dx.doi.org/10.1007/s11051-012-0990-6

13. Machunsky, S. and Peuker, U.A. (2007) Liquid-Liquid Interfacial Transport of Nanoparticles. Physical Separation in Science and Engineering, 2007, Article ID: 34832.http://dx.doi.org/10.1155/2007/34832

14. Youssef, A.A., Al-Dahhan, M.H. and Dudukovic, M.P. (2013) Bubble Columns with Internals: A Review. International Journal of Chemical Reactor Engineering, 11, 1-55.http://dx.doi.org/10.1515/ijcre-2012-0023

15. Shaikh, A. and Al-Dahhan, M. (2007) A Review on Flow Regime Transition in Bubble Columns. International Journal of Chemical Reactor Engineering, 5, 1-68.

16. Vecer, M., Lestinsky, P., Wichterle, K. and Ruzicka, M. (2012) On Bubble Rising in Countercurrent Flow. International Journal of Chemical Reactor Engineering, 10, 1-19.http://dx.doi.org/10.1515/1542-6580.2995

17. Hadavand, L. and Fadavi, A. (2013) Effect of Vibrating Sparger on Mass Transfer, Gas Holdup, and Bubble Size in a Bubble Column Reactor. International Journal of Chemical Reactor Engineering, 11, 1-10. http://dx.doi.org/10.1515/ijcre-2012-0094

18. Lai, R.W.M. and Fuerstenau, D.W. (1968) Liquid-Liquid Extraction of Ultrafine Particles. Transactions of the American Institute of Mining, Metallurgical, and Petroleum Engineers, 241, 549-556.

19. Erler, J., Machunsky, S., Grimm, P., Schmid, H.-J. and Peuker, U.A. (2013) Liquid-Liquid Phase Transfer of Magnetite Nanoparticles—Evaluation Of Surfactants. Powder Technology, 247, 265-269. http://dx.doi.org/10.1016/j.powtec.2012.09.047

20. Zhang, L., He, R. and Gu, H.-C. (2006) Oleic Acid Coating on the Monodisperse Magnetite Nanoparticles. Applied Surface Science, 253, 2611-2617.http://dx.doi.org/10.1016/j.apsusc.2006.05.023

21. Rudolph, M., Erler, J. and Peuker, U.A. (2012) A TGA-FTIR Perspective of Fatty Acid Adsorbed on Magnetite Nanoparticles—Decomposition Steps and Magnetite Reduction. Colloid and Surfaces A: Physicochemical and Engineering Aspects, 397, 16-23.http://dx.doi.org/10.1016/j.colsurfa.2012.01.020

22. Blaß, E. (1988) Bildung und Koaleszenz von Blasen und Tropfen. Chemie Ingenieur Technik, 60, 935-947. http://dx.doi.org/10.1002/cite.330601203

23. Räbiger, N. and Schlüter, M. (2006) Bildung und Bewegung von Tropfen und Blasen. In: VDI Wärmeatlas, Springer, Berlin, 1-15.

24. Stackelberg, M.V. (1949) Spontane Emulgierung Infolge Negative Grenzflächenspannung. Kolloid-Zeitschrift, 115, 53-66. http://dx.doi.org/10.1007/BF01501433

25. Kubatta, E.A. and Rehage, H. (2009) Characterization of Giant Vesicles Formed by Phase Transfer Processes. Colloid and Polymer Science, 287, 1117-1122.http://dx.doi.org/10.1007/s00396-009-2083-3

26. Levenspiel, O. (1999) Chemical Reaction Engineering. John Wiley & Sons, Inc., Hoboken.

27. Chapman, D.G., King, G.G., Berend, N., Diba, C. and Salome, C.M. (2010) Avoiding Deep Inspirations Increases the Maximal Response to Methacholine Without Altering Sensitivity in Non-Asthmatics. Respiratory Physiology and Neurobiology, 173, 157-163.http://dx.doi.org/10.1016/j.resp.2010.07.011

Thermodynamic Study for Arsenic Removal from Freshwater by Using Electrocoagulation Process

J. R. Parga[1], J. L. Valenzuela[2], G. T. Munive[2], V. M. Vazquez[2], and M. Rodriguez[1]

[1]Department of Materials Science, Technological Institute of Saltillo, Saltillo Coahuila, Mexico

[2]Department of Chemical Engineering and Metallurgy, University of Sonora, Hermosillo, Mexico

ABSTRACT

Insert Industrial treatment of mineral-processing and non-ferrous metal-smelting acid wastewater effluents is becoming an enormous worldwide problem. In Mexico, heavy metalscontaminated natural waters, including freshwater, surface water and ground water, are a significant problem as some of these compounds are known as toxic, mutagenic, and carcinogenic. A very promising electrochemical treatment technique that does not require chemical additions to remove arsenic is electrocoagulation (EC) with air injection. The proposed electrochemical process is efficient because used low cost iron electrodes and promising in industrial application. Theoretical the purpose of this research was to investigate the thermodynamic of arsenic adsorption on iron species using the Langmuir's Isotherm.

Also, thermodynamic parameters such as ΔH°, ΔS° and ΔG° were calculated and the adsorption process was found to be exothermic and spontaneous. X-ray Diffraction and Scanning Electron Microscopy, were used to characterize the solid products formed during EC. The results of this study suggest that magnetite particles and amorphous iron oxyhydroxides are present in the examined EC products and this study indicate that arsenic can be successfully adsorbed on iron species by electrocoagulation process. Field pilot-scale study demonstrated the removal of As(III)/As(V) with an efficiency of more than 99% from both wastewater and wells.

INTRODUCTION

Arsenic contaminated natural waters, including surface water and ground water, are a significant problem as some of these compounds are known to be toxic, mutagenic, and carcinogenic. The deleterious health effects associated with ingestion of arsenic require that its concentrations should be kept below 10.0 ppb in potable water supplies. The metals existing in wastewater are usually removed by precipitation. Arsenic is conventionally removed by chemical techniques such as treatment with lime, aluminum coagulation, iron coagulation, activated carbon and reverse osmosis [1] . These conventional processes generate a considerable quantity of secondary pollutants such as solid sludge, which also pose serious environmental problems. These drawbacks have forced various industries to search for effective alternative treatment technologies for arsenic removal, mainly electrochemical methods. The purpose of this research was to investigate the arsenic adsorption on iron species. A very promising electrochemical treatment technique that does not require addition of chemicals is electrocoagulation (EC) with air injection [2] . The EC process operates on the principle that the cations produced electrolytically from iron and/or aluminum anodes enhance the coagulation of contaminants from an aqueous medium. The sacrificial metal anodes are used to continuously produce polyvalent metal cations in the vicinity of the anode. These cations facilitate coagulation by neutralizing the negatively charged particles that are carried toward the anodes by electrophoretic motion. In the flowing EC techniques, the production of polyvalent cations from the oxidation of the sacrificial anodes (Fe and/or Al) and the production of electrolysis gases (H_2 and O_2) are directly proportional to the amount of current applied (Faraday's law) [1] . The electrolysis gases enhance the flocculation of the coagulant materials.

EC is a complicated process involving many chemical and physical phenomena. It can be said that in an EC process the coagulant is produced "in situ" and this involves three successive stages:

- Formation of coagulants by electrolytic oxidation of the "sacrificial electrode". Fe is dissolved from the anode generating corresponding metal ions, which almost immediately hydrolyze to ferric hydroxide and polymeric Iron; the generated $Fe^{3+}_{(aq)}$ ions will immediately undergo further spontaneous reactions to produce corresponding hydroxides and/or polyhydroxides. Ferric ions generated by electrochemical oxidation of iron electrode may form monomeric ions, $Fe(OH)_3$ and polymeric hydroxy complexes, namely: $Fe(H_2O)_6^{3+}$, $Fe(H_2O)_5(OH)_2^+$, $Fe(H_2O)_4(OH)^{2+}$, $Fe(H_2O)_8(OH)_2^{4+}$ and $Fe(H_2O)_6(OH)_4^{4+}$ depending on the pH of the aqueous medium. These hydroxides/polyhydroxides/ polyhydroxymetallic compounds have strong affinity for dispersed particles as well as counter ions to cause coagulation. The gas evolved at the cathode may impinge on and cause flotation of the coagulated materials.

- Destabilization of the contaminants, particulate suspension, and breaking of emulsions. Contaminants present in the wastewater stream are treated either by chemical reactions and precipitation or physical and chemical attachment to coagulant materials.

- Aggregation of the destabilized phases to form flocs. Polymeric hydroxides are excellent coagulating agent

EC has been employed for removing heavy metals such as Pb, Cd, Cr and As, metals such as Mn, Cu, Zn, Ni, Al, Fe, Co, Sn, Mg, Se, Mo, Ca, Pt among others, anions such as CN, PO_4, SO_4, NO_3, F and Cl, non metals such as P, organic compounds such as TPH, TBX, MTBE, COD,

BOD, suspended solids, clay minerals, organic dyes, oil and greases from a variety of industrial effluents. Some of the factors that influence EC performance are: type of electrolytes, electrodes material, applied power, acidity and final pH. A schematic representation of the EC process is shown in Figure 1.

The chemical reactions that have been proposed to describe EC mechanisms for the production of $H_{2(g)}$, $H^-_{(aq)}$ and $H^+_{(aq)}$ [3] [4] at different pH values are:

For pH < 4

Anode:

$$Fe \rightarrow Fe^{+2} + 2e$$

(1)

$$Fe \rightarrow Fe^{+3} + 3e$$

(2)

Cathode:

$$2H^+ + 2\,e^- \rightarrow H_{2(g)} \uparrow$$

(3)

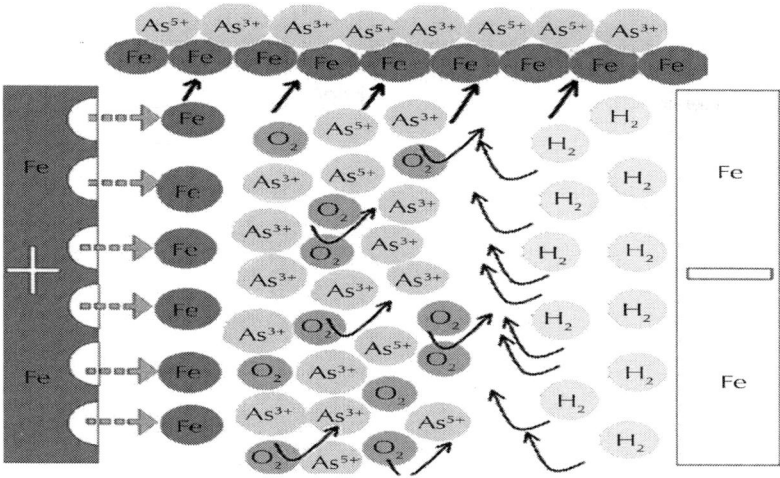

Figure 1: Conceptual illustration of the EC mechanism.

For 4 < pH < 7:

Anode: reactions (1) and (2)

Comments:

In fact iron also undergoes hydrolysis

$$Fe + 6H_2O \rightarrow Fe(H_2O)_4(OH)_2$$

(4)

$$Fe + 6H_2O \rightarrow Fe(H_2O)_3(OH)_{3(aq)} + 3H^{+1} + 3e^{-1}$$

(5)

Fe(III) hydroxide begins to precipitate floc with yellowish color.

$$Fe(H_2O)_3(OH)_{3(aq)} \rightarrow Fe(H_2O)_3(OH)_{3(s)}$$

(6)

"Rust" may also be formed [4] .

$$2\text{Fe}(\text{H}_2\text{O})_3(\text{OH})_3 \leftrightarrow \text{Fe}_2\text{O}_3(\text{H}_2\text{O})_6$$

(7)

Cathode: More hydrogen evolution takes place (Equation (3)); but $[\text{H}^+]$ now comes from weak acids and iron hydrolysis.

For $6 < \text{pH} < 9$:

Anode: reactions (1) and (2)

Comments:

Precipitation of Fe(III) hydroxide (7) continues, and Fe(II) hydroxide precipitation also occurs presenting a dark green floc.

$$\text{Fe}(\text{H}_2\text{O})_4(\text{OH})_{2(\text{aq})} \rightarrow \text{Fe}(\text{H}_2\text{O})_4(\text{OH})_{2(\text{s})}$$

(8)

The pH for minimum solubility of Fe(OH)_n is in the range of 7 - 8. EC floc is formed due to the polymerization of iron oxyhydroxides [5] .

Formation of rust (dehydrated hydroxides) occurs as shown in the following:

$$2\text{Fe}(\text{OH})_3 \leftrightarrow \text{Fe}_2\text{O}_3 + 3\text{H}_2\text{O} \ (\text{hematite, maghemite})$$

(9)

$$\text{Fe}(\text{OH})_2 \leftrightarrow \text{FeO} + \text{H}_2\text{O}$$

(10)

$$2\text{Fe}(\text{OH})_3 + \text{Fe}(\text{OH})_2 \leftrightarrow \text{Fe}_3\text{O}_4 + 4\text{H}_2\text{O} \ (\text{magnetite})$$

(11)

$$\text{Fe}(\text{OH})_3 \leftrightarrow \text{FeO}(\text{OH}) + \text{H}_2\text{O} \ (\text{goethite, lepidocrocite})$$

(12)

Hematite, maghemite, rust, magnetite, lepidocrocite and goethite have been identified as EC by-products [6] .

Cathode: More hydrogen evolution takes place (Equation (3)); but [H$^+$] now comes from weak acids and iron hydrolysis.

Overall reactions are:

$$Fe + 6H_2O \rightarrow Fe(H_2O)_4 (OH)_{2(s)} + H_{2(g)} \uparrow$$

(13)

$$Fe + 6H_2O \rightarrow Fe(H_2O)_3 (OH)_{3(s)} + \frac{3}{2} H_{2(g)} \uparrow$$

(14)

Conditions throughout the cell are not constant; concentrations, species and pH are changing. This may be illustrated with an iron Pourbaix diagram.

MATERIALS AND METHODS

The adsorption experiments were performed in a 400 ml glass beaker equipped with two carbon steel electrodes (6 cm × 3 cm) located 5 mm apart. There was used a source of current and voltage (universal AC/DC adaptor). The pH was measured with an electrode/pH meter-VWR scientific 8005. Arsenic adsorption onto iron species was investigated with electrolytes prepared with sodium arsenate Na_2HAsO_4, (analytical reagent with a purity of 97% supplied by Chemical Products, Monterrey) and deionized water with conductivity of 0.95 µS cm^{-1} (Aldrich Chemical Co. 99.5%). The solutions and solids were then separated by filtration through cellulose filter paper. The sludge from the EC was dried either in an oven or under vacuum at room temperature. The experimental set-up is presented in Figure 2. The current and voltage during the EC process were measured and recorded, using Cen-Tech multimeters. The pH values of the solution before and after EC were measured with a VWR scientific 8005 pH meter.

In Table 1, the initial conditions of arsenic for the range of concentrations listed are presented. The Langmuir isotherm model was used to analyze adsorption data.

The amount of arsenic adsorbed onto iron species at equilibrium N (mg/g) was calculated by using the following relationship [7] :

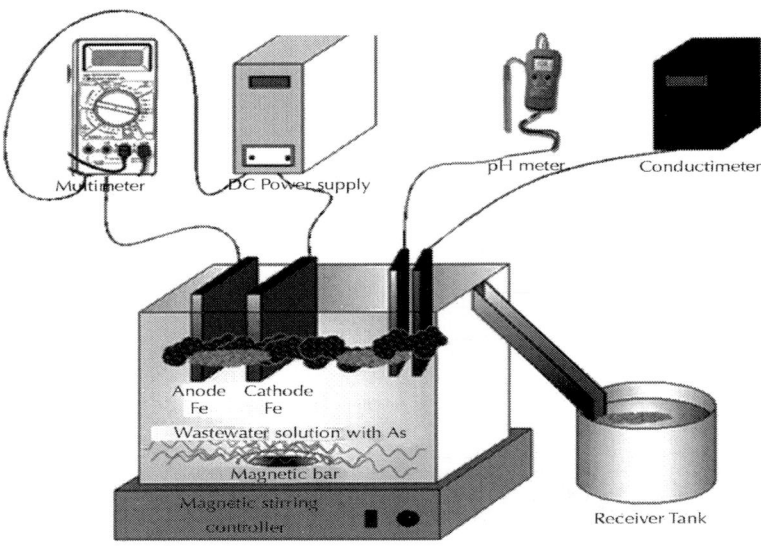

Figure 2: Experimental set-up for arsenic removal two-electrode electrocoagulation cell.

Table 1: Conditions for arsenic adsorption experiments

Samples	Concentration (mg/L)	Temperature (°C)	pH	Current (A)	Cell Potential (V)	Time (min)
1	1	20	2.86	0.49	8.5	5
2	2	20	3.02	0.52	12	5
3	5	20	2.90	0.44	10	5
4	7	20	3.05	0.41	12.8	5
5	13	20	2.90	0.48	10.7	5
6	20	20	3.02	0.50	10.25	5
7	30	20	4.14	0.75	6.68	5

$$N = V\left[C_0 - C_e\right]/W$$

(15)

which, C_0 and C_e (mg/L) are the liquid phase concentrations of arsenic initially and at equilibrium, respectively, V the volume of the solution (L) and W is the mass of adsorbent used (g) (in this case the Fe dissolved from EC). The mass of metal dissolved (or other species discharged) depends on the quantity of charge (electrons) supplied to the electrocoagulation cell during the holding time that happens across the electrolytic solution and the time of residence of the water in the electrocoagulation cell. A simple relation between the current density and the quantity of substance dissolved (g of M per cm²) stems from Faraday's law [7] :

$$W = \frac{D \cdot t \cdot M}{n \cdot F}$$

(16)

where W is the mass of the electrode dissolved (g/cm²), D is the density of current (A per cm⁻²), t is the time (seconds), M is the relative molar mass of the electrode, n the electrons number in the electrode reaction, and F is Faraday's constant (96,500 coulombs). Then, arsenic moles numbers adsorbed onto iron species are calculate by Equation (15).

In Table 2 arsenic moles number adsorbed onto iron species are presented. High adsorption was showed while the concentration increases.

Then, the Langmuir isotherm was employed to describe the adsorption equilibrium in the system:

$$N = \left(\frac{N_{Max}K_L C_e}{1 + K_L C_e}\right)$$

(17)

where N is the amount of arsenic adsorbed onto iron species in equilibrium (mg/g), N_{max} is the maximum adsorption capacity corresponding to complete monolayer coverage on the surface (mg/g), C_e is the equilibrium concentrations of metals ions in the solution at equilibrium (mg/L) and K_L is the Langmuir constant (L/mg). Equation (7) can be rearranged to a linear form as follows [8] :

$$\frac{C}{N} = \frac{1}{N_{Max} K_L} + \frac{C_e}{N_{Max}}$$

(18)

RESULTS AND DISCUSSION

Following the model of the Langmuir's isotherm (see Figure 2) shows the arsenic adsorption capacity on iron species. The constants can be evaluated from the intercepts and the slopes of the linear plots of C_e/N versus C_e. Table 2, shows the results of the linear regression. The value of N_{max} indicates the maximum adsorption capacity corresponding to complete monolayer coverage on the iron surface. The value of K_L is the constant of adsorption of Langmuir.

The thermodynamic parameters [such as change in standard free energy $\Delta G°$, enthalpy $\Delta H°$ and entropy $\Delta S°$ were determined [5] by using the following equations:

$$\ln b = -\frac{\Delta G°}{RT}$$

(19)

$$\ln\left(b\right) = b_0 - \left(\dfrac{\Delta H^\circ}{RT}\right)$$

$$(20)$$

Table 2: Arsenic moles adsorbed on iron species (N)

Sample	COAs (mmol/L)	CeAs (mmol/L)	W (g)	N (mmol/g)
1	0.013347	0.0002	0.028435	0.161827
2	0.026695	0.0006	0.033658	0.270655
3	0.066737	0.0009	0.025534	0.901994
4	0.093432	0.0020	0.023793	1.344981
5	0.173518	0.0049	0.027855	2.118223
6	0.266951	0.0126	0.029016	3.067147
7	0.400427	0.0168	0.043523	3.084857

$$\Delta G^\circ_{ads} = \Delta H^\circ_{ads} - T \Delta S^\circ_{ads}$$

$$(21)$$

In which R is the gas constant, T ($^\circ$K) the absolute temperature, b is the Langmuir's constant which is related to the energy of adsorption, and b_0 is a constant. Table 3; show the negative value of ΔG° obtained, indicated that the arsenic adsorption on iron species by EC is a spontaneous process. The negative value of ΔH° represents an exothermic process and suggests a fisiadsorption behavior according with the liberated heat between −20 and −40 KJ/mol. The negative value of ΔS° suggests that a significant change does not happen in the internal structure of the adsorbent during the adsorption of As.

PRODUCT CHARACTERIZATION

In this study, chemical analysis, powder X-ray diffraction (XRD), and scanning electron microscopy (SEM) were used to characterize the solid products generated in the EC cell, operated with carbon-steel electrodes. Figure 3 shows the diffractogram for an arsenic sample obtained from a solution at pH 7, containing 2 ppm arsenic. The species identified were magnetite, goethite, lepidocrocite, iron arsenate and iron hydroxide oxide in crystalline form.

Figure 4, shows, respectively, the SEM and EDAX images of As contaminated iron oxide/oxyhydroxide particles. These displays show that the surfaces of these iron oxide/oxyhydroxide particles were coated with a layer of As species.

Also, the results of this study suggest that EC produces magnetic particles of magnetite and amorphous iron oxyhydroxide species that serve to remove As(III)/As(V) species. In order to prove this, FTIR spectra for solids collected from EC trials with various starting pH value of ca. 7.0. The positions of the important absorption bands for the various iron containing phases are summarized in Table 1 [9] - [15] . This table is modeled after the work by Nauer et al. [9] that tabulated the peak positions of goethite (δ-FeOOH), akaganéite (δ-FeOOH), lepidocrocite (δ-FeOOH), feroxyhite (δ-FeOOH), and bernalite (Fe(OH)$_3$). These authors also used the follow- ing assignments for IR absorbances:

- 3600 to ca. 3000 cm^{-1}: OH stretching vibration, ν(OH).
- 1200 to 600 cm^{-1}: Fe-O-H bending vibrations and Fe-O stretching vibrations.
- 600 to 200 cm^{-1}: absorptions caused by the overlapping of lattice vibrations with molecular frequencies (Fe-O stretching vibrations can also absorb down to ca. 350 cm^{-1}).

After examining the spectra in that work, the current authors added the indications of shoulders on the ν(OH) stretching vibrations for δ-FeOOH, δ-FeOOH, and δ-FeOOH. For δ-FeOOH, trimodality of the ν(OH) stretching vibrations has also been indicated by adding the peaks at both 3450 and 2900 cm^{-1}. Another extensive study by Weckler and Lutz [12] found spectra for δ-FeOOH, δ-FeOOH, δ-FeOOH, and δ-FeOOH that closely resemble those summarized in Table 4. However, two additional features were added for goethite. The earlier work of Misawa et al. [13] also confirm the results discussed above for δ-FeOOH and also assigned a peak at 470 cm^{-1} and shoulders at 668 and 442 cm^{-1} to magnetite (Fe$_3$O$_4$). However, Musić [14] and Nasrazadani [10] observed peaks at ca. 575 and 385 cm^{-1} for magnetite and Nyquist and Kagel [15] observed peaks at 725 and 575 cm^{-1} for magnetite. The study by Musić [15] was also used as a reference for additional peaks for Fe(OH)$_3$, while Nasrazadani also assigned peaks at 630 and 430 cm^{-1} to maghemite (δ-Fe$_2$O$_3$).

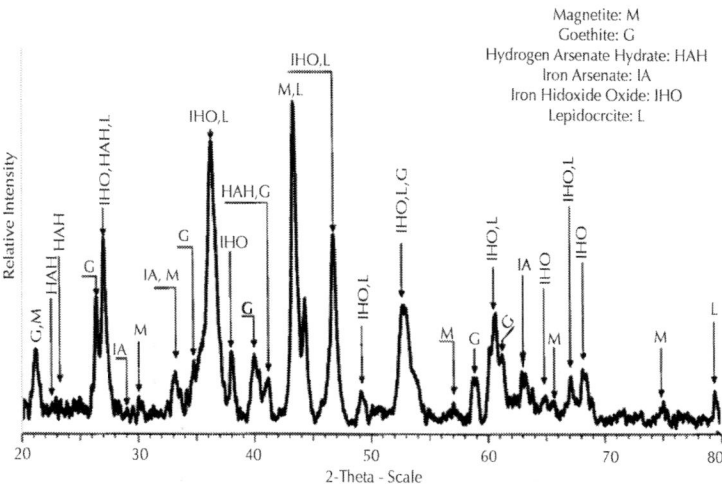

Figure 3: XRD diffractogram for the iron impregnated with arsenic.

Table 3. Linear regression results for arsenic adsorption onto iron species

Parameter	Results
R2	0.9665
Nmax (mmolAs/gFe)	4.242281
KL (L/mmol)	181.30

Table 4: Infrared frequencies of the Fe-oxyhydroxides (cm⁻¹) [9] -[15]

		Goethite α-FeOOH	Akaganéite β-FeOOH	Lepidocrocite γ-FeOOH	Feroxyhite δ-FeOOH	Bernalite Fe(OH)3	Hematite α-Fe2O3	Maghemite γ-Fe2O3	Magnetite Fe3O4
1)	sh	3500*	3480 s		3450* m	3500 - 2500			
	s	3155	3000* sh	3100 s	3150 s	1507c			
		1385d		2800* sh	2900* s				
2)	w	1125a		1155 m	1130 vw	1044c	1010f w		
	w	1081a		1160f s					
	w	1020f		1020 m	915 m				
				1020f s					
	s	890	875 sh				885f w		
	s	790	800 w	740 s	795 s				
			700 s				725e	785d	725e
	s	640	620 w				590e	630d sh	668b
							535e		
3)	sh	455		480 ss	460 ss	566c	568c	550d s	570c,d s
						448c			470b
	ss	400	425 ss	360 ss					442b sh
	s	270		270 ss	335 ss				385c,d

Peak notes: very strong (ss), strong (s), medium (m), weak (w), very weak (vw), shoulder (sh), broad (b).

CONCLUSIONS

The results of this study indicate that arsenic is readily adsorbed onto iron species generated by electrocoagula- tion. It was found that 99% of the arsenic in a batch cell is removed in the lab-scale EC reactor within 60 seconds or less for most experiments, and with a current efficiency of 100%. Correlation of the adsorption behavior with the Langmuir isotherm revealed an adsorption capacity of 4.24 mmol As/g of Fe absorbent. The negative values of $\Delta G°$ obtained indicate that the arsenic adsorption onto iron species generated by EC is a spontaneous process. The negative values of $\Delta H°$ represent an exothermic process.

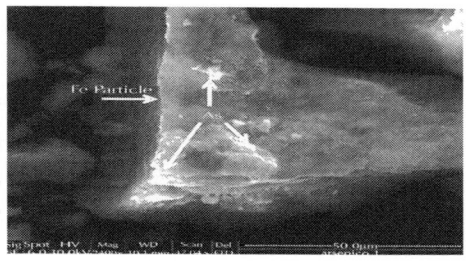

(a)

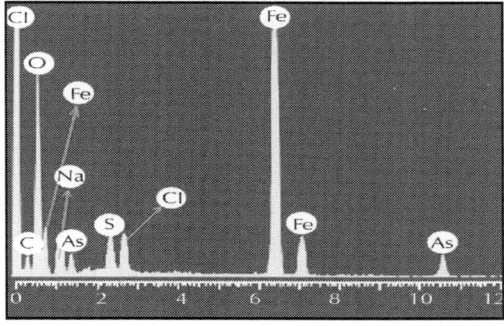

(b)

Figure 4: SEM and EDAX images for by-products generated from the arsenic contaminated water samples.

ACKNOWLEDGEMENTS

The authors acknowledge the financial support of this research provided by CONACYT, TecNM (ITS), USAID and GOBIERNO DE COAHUILA for his contribution in the presentation of these manuscript.

REFERENCES

1.   Gomes, J.A., Cocke, D.L., Daida, P., Kesmez, M., Weir, M., Moreno, H., Parga, J.R., Irwin, G., McWhinney, H., Gradyand, T. and Peterson, E. (2007) Arsenic Removal by Electrocoagulation Using Combined Al-Fe Electrode System and Characterization of Products. Journal of Hazardous Material, B139, 220-231.http://dx.doi.org/10.1016/j.jhazmat.2005.11.108

2.   Parga, J.R., Cocke, D.L., Valenzuela, J.L., Gomes, J.A., Kesmez, M., Irwin, G., Moreno, H. and Weir, M. (2005) Arsenic Removal via Electrocoagulation from Heavy Metal Contaminated Groundwater in La Comarca Lagunera Mexico. Journal of Hazardous Materials, B124, 247-254. http://dx.doi.org/10.1016/j.jhazmat.2005.05.017

3.   Mohora, E., Roncevic, S., Agbaba, J., Tubic, A. and Mitic, M. (2014) Removal of Arsenic from Groundwater Rich in Natural Organic Matter (NOM) by Continuous Electrocoagulation/Flocculation (ECF), Separation and Purification Technology, Available online 16 September 2014.

4.   Moreno, H., Cocke, D.L., Gomes, J.A., Morkovsky, P., Parga, J.R., Peterson, E. and Garcia, C. (2007) Electrochemical Generation of Green Rust with Electrocoagulation. ECS Transactions, 3, 67-76.

5.   Parga, J.R., Vazquez, V. and Casillas, H. (2009) Cyanide Detoxification of Mining Wastewaters with TiO_2 Nanoparticles and Its Recovery by Electrocoagulation. Chemical Engineering and Technology, 32, 1901-1908. http://dx.doi.org/10.1002/ceat.200900177

6.   Roy, P., Mondale, N.K. and Das, K. (2014) Modeling of the Adsorptive removal of Arsenic: A Statistical Approach. Journal of Environmental Chemical Engineering, 2, 585-597.

7. Mólgoraa, C.C., Domíngueza, A.M., Avilab, E.M., Droguic, P. and Buelnad, G. (2013) Removal of Arsenic from Drinkingwater: A Comparative Study between Electrocoagulation-Micrifiltration and Chemical Coagulation-Microfiltration Process. Separation and Purification Technology, 118, 645-651.http://dx.doi.org/10.1016/j.seppur.2013.08.011

8. Zhang, P., Tong, M., Young, S. and Liao, P. (2014) Transformation and Removal of Arsenic in Groundwater by Sequential Anodic Oxidation and Electrocoagulation. Journal of Contaminant Hydrology, 164, 299-307. http://dx.doi.org/10.1016/j.jconhyd.2014.06.009

9. Nauer, G., Stretcha, P., Brinda-Konopik, N. and Liptay, G. (1985) Spectroscopic and Thermoanalytical Characte- rization of Standard Substances for the Identification of Reaction Products on Iron Electrodes. Journal of Thermal Analysis, 30, 813-830. http://dx.doi.org/10.1007/BF01913309

10. Nasrazadan, S. and Raman, A. (1993) The Application of Infrared Spectroscopy to the Study of Rust Systems—II. Study of Cation Deficiency in Magnetite (Fe_3O_4) Produced during Its Transformation to Maghemite (-Fe_2O_3) and Hematite (-Fe_2O_3). Corrosion Science, 34, 1355-1365. http://dx.doi.org/10.1016/0010-938X(93)90092-U

11. Parfitt, R.L. and Smart, R. (1978) The Mechanism of Sulfate Adsorption on Iron Oxides[1]. Soil Science Society of American Journal, 42, 48.http://dx.doi.org/10.2136/sssaj1978.03615995004200010011x

12. Weckler, B. and Lutz, H.D. (1998) Lattice Vibration Spectra. Part XCV. Infrared Spectroscopic Studies on the Iron Oxide Hydroxides Goethite ( ), Akaganéite ( ), Lepidocrocite ( ) and Feroxyhite ( ). European Journal of Solid State and Inorganic Chemistry, 35, 531-544. http://dx.doi.org/10.1016/S0992-4361(99)80017-4

13. Misawa, T., Kyuno, T., Suëtaka, W. and Shimodaira, S. (1971) The Mechanism of Atmospheric Rusting and the Effect of Cu and P on the Rust Formation of Low Alloy Steels. Corrosion Science, 11, 35-48. http://dx.doi.org/10.1016/S0010-938X(71)80072-0

14. Musi , S., Goti , M. and Popovi , S. (1993) X-Ray Diffraction and Fourier Transform-Infrared Analysis of the Rust Formed by

Corrosion of Steel in Aqueous Solutions. Journal of Materials Science, 28, 5744-5752. http://dx.doi.org/10.1007/BF00365176

15.   Nyquist, R.A. and Kagel, R.O. (1971) Infrared Spectra of Inorganic Compounds (3800-450 cm^{-1}), 218. Academic Press, New York and London.

Chapter 7

Design of a Simulator for Enhanced Oil Recovery Process Using a Nigerian Reservoir as a Case Study

Kamilu Folorunsho Oyedeko[1]
and Alfred Akpoveta Susu[2]

[1]Department of Chemical & Polymer Engineering, Lagos State University, Epe, Lagos, Nigeria

[2]Department of Chemical Engineering, University of Lagos, Lagos, Nigeria

ABSTRACT

This study involves the applications of different numerical techniques in a more general way to the design of a simulator for an enhanced oil recovery process with surfactant assisted water flooding. The data from a hypothetical oil well and a Nigerian oil well were used for the validation of the simulator. The process is represented by a system of nonlinear partial differential equations: the continuity equation for the transport of the components and Darcy's equation for the phase flow. The orthogonal collocation, finite difference and coherence theory techniques were used in solving the equations that characterized the multidimensional, multiphase and multicomponent flow problem. Matlab computer programs were used for the numerical solution of the model equ- ations. The predicted simulator, obtained from the resulting numerical exercise confers uncondi- tional stability and more insight into the physical reservoir description. The results of the ortho- gonal collocation solution were compared with those of finite difference and coherence solutions. The results indicate that the concentration of surfactants for orthogonal collocation show more features than the predictions of the coherence solution and the finite difference, offering more opportunities for further understanding of the physical nature of the complex problem. We have found out that the partition of the three components between the two-phases determines other physical property data and hence the oil recovery. The oil recovery for the Nigerian oil reservoir is higher than the recovery predicted for the hypothetical crude. The displacement mechanism for the multicomponent and multiphase system is stable in the Nigerian oil reservoir due to the mod- erate value of the oil/water viscosity instead of the hypothetical reservoir with high value of oil/water ratio.

INTRODUCTION

The world economy today can be characterized as a crude oil economy. So far, there has not been a single energy source that has broadly been integrated to replace crude oil in the provision of electricity (light and heat) and transportation (land, air and sea). It is very important to at least, maintain or indeed, increase the current production levels of crude oil. These objectives can be accomplished by further investing in exploration and production of new fields or optimizing production from existing fields. Bringing new fields online can be expensive, while recovery from existing fields by conventional methods (i.e. primary and secondary recovery) will not fully provide the necessary relief for global oil demand.

On an average, only about a third of the original oil in place can be recovered by primary and secondary recovery processes. The rest of the oil is trapped in reservoir pores due to surface and interfacial forces. This trapped oil can be recovered by reducing the capillary forces that prevent oil from flowing within the pores of reservoir rock and into the well bores. Due to high oil prices and declining production in many regions around the globe, the application of advance technologies called "Enhanced Oil Recovery" (EOR) has become very attractive for exploration and production of the trapped oil. This technology requires the injection of a fluid or fluids or materials into a reservoir to supplement the natural energy present in a reservoir, where the injected fluids interact with the reservoir rock/oil/brine system to create favourable conditions for maximum oil recovery. Surfactants are injected to decrease the interfacial tension between oil and water in order to mobilize the oil trapped after secondary recovery by water flooding.

In a surfactant flood, a multi-component multiphase system is involved. The theory of multi-component, multiphase flow has been presented by several authors. The surfactant flooding is a form of chemical flooding and is represented by a system of nonlinear partial differential equations: the continuity equation for the transport of the components and Darcy's equation for the phase flow. The system of equations is completed by the equations representing physical properties of the fluids and the rock. From a physico-chemical point of view, there are three components—water, petroleum and chemical. They are in fact, pseudo-components, since each one consists of several pure components. Petroleum is a complex mixture of many hydrocarbons. Water is actually brine, and contains dissolved salts. Finally, the chemical contains different kinds of surfactants. These components are distributed between two phases—the oleic phase and the aqueous phase. The chemical has an amphiphilic character. It makes the oleic phase at least partially miscible with water or the aqueous phase, partially miscible with petroleum.

Interfacial tension depends on the surfactant partition between the two phases. Residual phase saturation decreases as interfacial tension decreases. Relative permeability parameters depend on residual phase saturations. In addition, phase viscosities are functions of the volume fraction of the components in each fluid phase. Therefore, the success or failure of surfactant flooding processes depends on phase behaviour.

Phase behaviour influences all other physical properties, and each of them, in turn influences oil recovery.

The different mathematical techniques used here, orthogonal collocation method, finite difference and coherence theory methods, are utilized for identification of particular physical behaviour. Besides, they may enable the understanding of the involved propagation phenomena in terms of cause and effects. More so, the techniques will in particular be utilized to predict what happens in EOR process and show how the complexity of the problem can be reduced by intensive calculation.

Systems of coupled, first-order, nonlinear hyperbolic partial differential equations (p.d.e.s) govern the transient evolution of a chemical flooding process for enhanced oil recovery. The method of characteristics (MOC) provides a way in which such systems of hyperbolic p.d.e.s can be solved by converting them to an equivalent system of ordinary differential equations. In fact, the MOC provides a similar framework for algorithm development to the coherence approach which is one of the numerical methods adopted here. In some cases, the characteristic solution has been used to track the flood-front in two-dimensional reservoir problems [1] . While the approach proposed by Ewing et al. [2] , combines the characteristic method with a finite element approach, Zheng [3] used the MOC and an adjustable number of moving particles to track three-dimensional solute fronts in groundwater systems by adjusting the number of particles served to maintain an accurate material balance and save computational time. This front-tracking approach has been used in the present work to trace the movement of coherent waves, of both the diffuse and shock variety.

The concept of coherence was extended to general EOR processes [4] [5] , including alkaline flooding [6] , convectional surfactant flooding [7] the effects of cation-exchange on surfactant-polymer flooding [8] [9] and miscible gas injection processes such as CO_2 flooding (Orr et al., 1995; Wang and Orr, 1997). Refinements to the theory also allowed for equilibrium reaction to occur, such as precipitation-dissolution [10] [11] and micelle formation [12] . Helset and Lake [13] have used simple wave theory (essentially identical to coherence theory) to study the one dimensional, three phase secondary migration of hydrocarbons from a source rock into possible reservoirs. Most recently, the theory of coherence has been applied successfully to the analysis of the transport of volatile compounds in porous media in the presence of a trapped gas phase [14] .

At the simple level, the results of simulation using the principle of coherence are analogous to the Buckley-Leverett theory for water flooding, the latter being evident in the work of Patton et al. [15] for polymer flooding, Fayers and Perrine [16] for dilute surfactant flooding, Claridge and Bondor [17] for carbonated water flooding and Larson [18] and Hirasaki [7] for miscible and immiscible surfactant flooding, respectively. Pope, et al. [19] for isothermal, multiphase, multicomponent fluid Flow in permeable media. Hankins and Harwell [20] . Case studies for the feasibility of sweep improvement in surfactant-assisted water flooding. Other works on EOR researches include the work of Siggel, et al. [21] for a new class of viscoelastic surfactants for EOR, Xu and Lu [22] for microbially enhanced oil recovery at simulated reservoir conditions by use of engineered bacteria, Andrew Leach and Mason [23] for co-optimization of enhanced oil recovery and carbon sequestration, Harwell [24] for development of improved surfactants and EOR methods for small operators and many others.

So, in this work, we apply the different numerical techniques, orthogonal collocation, finite difference and coherence theory method, in solving the basic model transport equations characterizing the design of the simulator. As far as the authors are aware, this is the first time that the orthogonal collocation method is being applied to simulator design. The approach is multidimensional and involves at least three independent variables with the introduction of the concept of partial coherence so that the various composition path spaces required for mapping the composition routes of the system are at most two dimensional.

METHODOLOGY

In this work, we considered solving multidimensional, multicomponent, multiphase flow problems associated with enhanced oil recovery process in petroleum engineering. The process of interest involves the injection of surfactant of different concentrations and pore volume to displace oil from the reservoir.

The methodology indicates the steps utilized in executing the project using the developed mathematical models to describe the physics of reservoir depletion and fluid flow in which one of the main aims is the determination of the areal distribution of fluids in the reservoir resulting from a flood. The system is for two or three dimensions, two fluid phases (aqueous, oleic) and one adsorbent phase, four components (oil, water, surfactants 1 and 2).

The reservoir may be divided into discrete grid blocks which may each be characterized by having different reservoir properties. The flow of fluids from a block is governed by the principle of mass conservation coupled with Darcy's law. The simultaneous flow of oil, gas, and water, in three dimensions and the effects of natural water influx, fluid compressibility, mass transfer between gas and liquid phases and the variation of such parameters as porosity and permeability, as functions of pressure are then modelled.

The model is developed from the basic law of conservation of mass [25] . The developed partial differential equation is converted to ordinary differential equation using coherent, finite difference and orthogonal collocation methods.

The finite difference method is a technique that converts partial differential equations into a system of linear equations. There are essentially three finite difference techniques. The explicit, finite difference method converts the partial differential equations into an algebraic equation which can be solved by stepping forward (forward difference), backward (backward difference) or centrally (central difference).

The orthogonal collocation method converts partial differential equations into a system of ordinary differential equations using the lagrangian polynomial method. The set of ordinary differential equations generated is then solved with appropriate numerical technique, in this work, by the explicit Runge Kutta method. The motivation for this explicit method is the simplicity and computation and efficiency and possibility of reducing truncation error over the required for other methods.

The coherent theory method combines the forward finite difference method with Runge Kutta technique to solve partial differential equations.

The rock and fluid properties such as density, porosity, viscosity, oil and water etc, and other parameters are listed in Tables 1-4. Table 1 is the Reservoir characteristics from the work by Hankins and Harwell

[25] . Table 2 is the Reservoir Characteristics used for the Simulation work by Oyedeko [26] . Parameter values used in Trogus adsorption model for verification runs is shown inTable 3 while Table 4 is the Additional Reservoir Parameters for the coherence work by Hankins and Harwell [25] .

The model encompasses two fluid phases (aqueous and oleic), one adsorbent phase (rock), and four components (oil, water, surfactants 1 and 2). The oil is displaced by water flooding. In-situ interaction of surfactant slugs may occur, with consequent phase separation and local permeability reduction. The model accommodates two (or three) physical dimensions, and an arbitrary, nonisotropic description of absolute permeability variation and porosity.

For most of the simulated cases in the work of Harkins and Harwell [25] , the reservoir consisted of a rectangular composite of horizontal oil bearing strata, sandwiched above and below by two impervious rocks. Oil is produced from the reservoir by means of water injection at one end and a production well at the other. Data for the hypothetical reservoir simulated in Hankins and Harwell [25] are given in Table 6 and the model developed is given as:

Table 1: Reservoir characteristics from the work of Hankins and Harwell [25]

Parameter	Value
Rock density	2.65 g/cm^3
Porosity	0.2
Oil viscosity	5.0 cp
Water viscosity	1.0 cp
Injection pressure gradient (maintained constant)	1.5 psi/ft
Fluid densities	1.0 g/cm^3
Width of injection face	50 ft
Width of central high permeability streak	10 ft
Length of reservoir	100 or 5000 ft
Residual oil saturation	0.2

Connate water saturation	0.1
First injected surfactant	SDS
Second injected surfactant	DPC
Henry's law constant SDS DPC	2.71×10^{-4} l/g 8.30×10^{-5} l/g
CMC values SDS DPC	800 µmol/l 4000 µmoll/l
Injected concentration SDS DPC	10 CMC 10 CMC
Brine spacer (typical)	≈0.05 pore volumes
Slug volumes	≈0.10 pore volumes

Table 2: Reservoir characteristics used for the simulation work by Oyedeko [26]

Parameter	Value
Rock density	2.65 g/cm^3
Porosity	0.2
Oil viscosity	0.40 cp
Water viscosity	0.30 cp
Injection pressure gradient (maintained constant)	1.5 psi/ft
Fluid densities	1.0 g/cm^3
Width of injection face	50 ft
Width of central high permeability streak	10 ft
Length of reservoir	100 or 5000 ft
Residual oil saturation	0.2
Connate water saturation	0.2
First injected surfactant	SDS
Second injected surfactant	DPC
Henry's law constant SDS DPC	2.71×10^{-4} l/g 8.30×10^{-5} l/g

CMC values SDS DPC	800 µmol/l 4000 µmoll/l
Injected concentration SDS DPC	10 CMC 10 CMC
Brine spacer (typical)	≈0.05 pore volumes
Slug volumes	≈ 0.10 pore volumes

Table 3: Parameter values used in Trogus adsorption model for verification runs

Parameter	Value
Pure component CMCs	$C_1^* = 1.0$ mol m^3 $C_2^* = 0.35$/mol/m^3
Phase separation model parameter	$= 1.8$
Henry's law constants for adsorption	$\overline{C_i} = k_i C_{i,w}$ ($C_{i,w}$ = aqueous monomer concentration) $k_1 = 0.21 \times 10^{-3}$ m^3/kg $k_2 = 0.80 \times 10^{-3}$ m^3/kg
Henry's law constant for oleic partitioning	$C_{i,o} = q_i C_{i,w}$ ($C_{i,w}$ = aqueous monomer concentration) $q_1 = 7.1$ $q_2 = 1.3$
Adsorbent properties	$_s = 2.1 \times 10^{+3}$ m^3/kg $= 0.2$

Table 4: Additional reservoir parameters for the coherence work by Hankin and Harwell [25]

Model designation	A	B
Grid points in the horizontal direction (m + 1)	21	21
Grid points in the vertical direction (n + 1)	11	21
Coherent waves of water saturation	28	28
Initial number of points per coherent wave Water Surfactant	41 81	41 81
Maximum number of points required per coherent wave	≈300	≈300
Average time step size (days) Short reservoir (100 ft) 200 mD streak 1000 mD streak Long reservoir (5000 ft) 200 mD streak 1000 mD sreak	3.47 0.69 174.0 34.7	3.47 0.69 174.0 34.7
Typical number of time steps required to inject first pore volume Short reservoir Long reservoir	33 75	33 75

$$\phi S_w \frac{\partial C_{i,w}}{\partial t} + \rho (1-\phi) \frac{\partial \overline{C_i}}{\partial t} + \phi v_x f_w \frac{\partial C_{i,w}}{\partial x} + \phi v_y f_w \frac{\partial C_{i,w}}{\partial y} = -r_i \quad (i=1,2)$$

(1)

The term r_i represents the rate of loss of surfactant due to precipitation: for a one-to-one reaction stoichiometry, $r_1 = r_2$. Since reaction occurs instantaneously at a sharp interface, this term may be ignored away from the singular region of the interface.

It is possible to approximate the adsorption isotherm of a pure surfactant on a mineral oxide by use of a simple model. At low concentration the adsorption obeys Henry's law, while above the critical micelle concentration (CMC) the total adsorption remains constant. The Trogus adsorption model [12] [27] is used in this work.

Application of Coherence Theory to the Solution of Model Equations

The material balance equations, Equation (1) (in the absence of r_i), are first order, homogeneous, nonlinear hyperbolic equations. Their solution will be attempted by means of the theory of coherence. The results presented here are general, and not restricted to assumptions regarding equilibrium relationships, fractional flow relationships, etc.

The concept of coherence identifies the state which a dynamic, multi-component system strives to attain. The state of "coherence" requires all dependent variables at any given point in space and time to have the same wave velocity, giving rise to "a coherent" wave with no relative shift in the profiles of the variables. It has been established mathematically by Helfferich [28] that an arbitrary starting variation of dependent variables, if embedded between sufficiently large regions of constant state, is resolved into coherent waves, which become separated by new regions of constant state.

Oleic Partitioning

The model developed and expressed in Equation (1) may be generalized by allowing surfactants to partition into the oleic phase. In general, $C_{i,o}=C_{i,o}(C_{1,w},C_{2,w})$ this leads to:

$$\phi S_w \frac{\partial C_{1,w}}{\partial t} + \phi S_o \frac{\partial C_{i,o}}{\partial t} + (1-\phi)\rho \frac{\partial \bar{C}_{1,w}}{\partial t} + \phi v_x f_w \frac{\partial C_{1,w}}{\partial x} + \phi v_y f_w \frac{\partial C_{1,w}}{\partial y} + \phi v_x f_o \frac{\partial C_{i,o}}{\partial x} + \phi v_y f_o \frac{\partial C_{i,o}}{\partial x} = -r_i \tag{2}$$

Leading to the matrix equation:

$$\begin{bmatrix} (S_w + S_o p_{11} + m_{11})\lambda - f_w - f_o p_{11} & (S_o p_{12} + m_{12})\lambda - f_o p_{12} \\ (S_o p_{21} + m_{21})\lambda - f_o p_{21} & (S_w + S_o p_{22} + m_{22})\lambda - f_w - f_o p_{22} \end{bmatrix} \times \begin{bmatrix} dC_{1,w} \\ dC_{2,w} \end{bmatrix} = 0 \tag{3}$$

where

$$p_{i,j} = \frac{\partial C_{i,o}}{\partial C_{j,w}}, \quad f_o = 1 - f_w, \quad S_o = 1 - S_w$$

Application of Finite Difference to the Solution of Model Equations

First-order, finite-difference expressions for the spatial derivatives were substituted into the hyperbolic chromatographic transport equations (Equation (1)), yielding 2× m coupled ordinary differential equations which may then be integrated simultaneously (also known as the "numerical method of lines").

$$s_w \frac{\partial C_{i,w}}{\partial \tau} + \sum_{j=1}^{2} m_{ij} \frac{\partial C_{i,w}}{\partial \tau} + f_w(\tau, \varepsilon_h) \times \left\{ \frac{C_{i,w}(\tau, \varepsilon_h) - C_{i,w}(\tau, \varepsilon_{h-1})}{\Delta \varepsilon} \right\} = 0 \tag{4}$$

where i=1,2 and $h = 1,2,\ldots,m$.

Equation (4) is the finite-difference form of Equation (1) written for one spatial dimension , where m_{ij} are the adsorption coefficients, τ is dimensionless time (injected volume/pore volume), and   is dimensionless distance (pore volumes travelled). In two dimensions, the finite-difference terms are multiplied by dimensionless velocities. The distortion of the solution in the τ direction may be neglected by using a 4th order Runge-Kutta method and a sufficiently small time step.

The above equation is now transformed to the original form of Equation (1) using the already defined variables below

$$C'_{i,w} = \phi C_{i,w} \tag{5}$$

$$\overline{C'_i} = \rho(1-\phi)\overline{C'_i} \tag{6}$$

$$m_{i,j} = \frac{\partial \overline{C'_i}}{\partial C'_{j,w}} \tag{7}$$

Again, recall that differentiation of a function of another function (chain rule) is of the form

$$\frac{\partial y}{\partial x} = \frac{\partial y}{\partial u} \times \frac{\partial u}{\partial x} \tag{8}$$

Applying the chain rule above, Equation (4) becomes

$$S_w \frac{\partial C'_{1,w}}{\partial \tau} + \left(\frac{\partial \overline{C'_1}}{\partial C'_{1,w}} \cdot \frac{\partial C'_{1,w}}{\partial \tau} \right) + \left(\frac{\partial \overline{C'_1}}{\partial C'_{2,w}} \cdot \frac{\partial C'_{2,w}}{\partial \tau} \right) + f_w\left(\tau, \varepsilon_h\right) \times \left\{ \frac{C'_{1,w}\left(\tau, \varepsilon_h\right) - C'_{1,w}\left(\tau, \varepsilon_{h-1}\right)}{\Delta \varepsilon} \right\} = 0 \tag{9}$$

Eliminating the primes (') and bars (−) and introducing $m_{i,j}$ terms yield

$$\left(S_w + m_{11}\right) \frac{\partial C_{1,w}}{\partial \tau} + m_{12} \frac{\partial C_{2,w}}{\partial \tau} + f_w \frac{\partial C_{1,w}}{\partial \varepsilon} = 0 \tag{10}$$

$$\left(S_w + m_{22}\right) \frac{\partial C_{2,w}}{\partial \tau} + m_{21} \frac{\partial C_{1,w}}{\partial \tau} + f_w \frac{\partial C_{2,w}}{\partial \varepsilon} = 0 \tag{11}$$

Applying the method of lines, a partial transformation to a difference equation, to the equations above yield:

$$\left(S_w + m_{11}\right) \frac{\partial C_{1,w}}{\partial \tau} + m_{12} \frac{\partial C_{2,w}}{\partial \tau} + f_w \frac{C_{1,w_{(\tau,\varepsilon_h)}} - C_{1,w_{(\tau,\varepsilon_{h-1})}}}{\Delta \varepsilon} = 0 \tag{12}$$

$$\left(S_w + m_{22}\right) \frac{\partial C_{2,w}}{\partial \tau} + m_{21} \frac{\partial C_{1,w}}{\partial \tau} + f_w \frac{C_{2,w_{(\tau,\varepsilon_h)}} - C_{2,w_{(\tau,\varepsilon_{h-1})}}}{\Delta \varepsilon} = 0 \tag{13}$$

This can also be written as follows:

$$\left(S_w + m_{11}\right) \frac{\partial C_{1,w_{(\tau,\varepsilon_h)}}}{\partial \tau} + m_{12} \frac{\partial C_{2,w_{(\tau,\varepsilon_h)}}}{\partial \tau} + \frac{f_w}{\Delta \varepsilon}\left[C_{1,w_{(\tau,\varepsilon_h)}} - C_{1,w_{(\tau,\varepsilon_{h-1})}} \right] = 0 \tag{14}$$

$$\left(S_w + m_{22}\right) \frac{\partial C_{2,w_{(\tau,\varepsilon_h)}}}{\partial \tau} + m_{21} \frac{\partial C_{1,w_{(\tau,\varepsilon_h)}}}{\partial \tau} + \frac{f_w}{\Delta \varepsilon}\left[C_{2,w_{(\tau,\varepsilon_h)}} - C_{2,w_{(\tau,\varepsilon_{h-1})}} \right] = 0 \tag{15}$$

Since we have a set of simultaneous ODE's, we will attempt to solve the equations:

$$\left(S_w + m_{11}\right) \frac{\partial C_{1,w_{(\tau,\varepsilon_h)}}}{\partial \tau} + m_{12} \frac{\partial C_{2,w_{(\tau,\varepsilon_h)}}}{\partial \tau} + \frac{f_w}{\Delta \varepsilon}\left[C_{1,w_{(\tau,\varepsilon_h)}} - C_{1,w_{(\tau,\varepsilon_{h-1})}} \right] = 0 \tag{16}$$

$$\left(S_w + m_{22}\right) \frac{\partial C_{2,w_{(\tau,\varepsilon_h)}}}{\partial \tau} + m_{21} \frac{\partial C_{1,w_{(\tau,\varepsilon_h)}}}{\partial \tau} + \frac{f_w}{\Delta \varepsilon}\left[C_{2,w_{(\tau,\varepsilon_h)}} - C_{2,w_{(\tau,\varepsilon_{h-1})}} \right] = 0 \tag{17}$$

Where

$$m_{11} = \frac{\partial \overline{C}_1}{\partial C_{1,w}}, \quad m_{12} = \frac{\partial \overline{C}_1}{\partial C_{2,w}}, \quad m_{21} = \frac{\partial \overline{C}_2}{\partial C_{1,w}}, \quad m_{22} = \frac{\partial \overline{C}_2}{\partial C_{2,w}} \tag{18}$$

Substituting for these terms in Equations (16) and (17) yield

$$\left(S_w + \frac{\partial \overline{C}_2}{\partial C_{2,w}} \right) \frac{\partial C_{2,w_{(\tau,\varepsilon_h)}}}{\partial \tau} + \frac{\partial \overline{C}_2}{\partial C_{1,w}} \frac{\partial C_{1,w_{(\tau,\varepsilon_h)}}}{\partial \tau} + \frac{f_w}{\Delta \varepsilon} \left[C_{2,w_{(\tau,\varepsilon_h)}} - C_{1,w_{(\tau,\varepsilon_h-1)}} \right] = 0 \tag{19}$$

And

$$\left(S_w + \frac{\partial \overline{C}_2}{\partial C_{2,w}} \right) \frac{\partial C_{2,w_{(\tau,\varepsilon_h)}}}{\partial \tau} + \frac{\partial \overline{C}_2}{\partial C_{1,w}} \frac{\partial C_{1,w_{(\tau,\varepsilon_h)}}}{\partial \tau} + \frac{f_w}{\Delta \varepsilon} \left[C_{2,w_{(\tau,\varepsilon_h)}} - C_{2,w_{(\tau,\varepsilon_h-1)}} \right] = 0 \tag{20}$$

These on simplification yield

$$S_w \frac{\partial C_{1,w_{(\tau,\varepsilon_h)}}}{\partial \tau} + \frac{\partial \overline{C}_1}{\partial C_{1,w}} \cdot \frac{\partial C_{1,w_{(\tau,\varepsilon_h)}}}{\partial \tau} + \frac{\partial \overline{C}_1}{\partial C_{2,w}} \cdot \frac{\partial C_{2,w_{(\tau,\varepsilon_h)}}}{\partial \tau} + \frac{f_w}{\Delta \varepsilon} \left[C_{1,w_{(\tau,\varepsilon_h)}} - C_{1,w_{(\tau,\varepsilon_h-1)}} \right] = 0$$

$$S_w \frac{\partial C_{1,w_{(\tau,\varepsilon_h)}}}{\partial \tau} + \frac{\partial \overline{C}_1}{\partial \tau} + \frac{\partial \overline{C}_1}{\partial \tau} + \frac{f_w}{\Delta \varepsilon} \left[C_{1,w_{(\tau,\varepsilon_h)}} - C_{1,w_{(\tau,\varepsilon_h-1)}} \right] = 0$$

$$S_w \frac{\partial C_{1,w_{(\tau,\varepsilon_h)}}}{\partial \tau} + 2 \frac{\partial \overline{C}_1}{\partial \tau} + \frac{f_w}{\Delta \varepsilon} \left[C_{1,w_{(\tau,\varepsilon_h)}} - C_{1,w_{(\tau,\varepsilon_h-1)}} \right] = 0$$

similarly

$$S_w \frac{\partial C_{2,w_{(\tau,\varepsilon_h)}}}{\partial \tau} + 2 \frac{\partial \overline{C}_2}{\partial \tau} + \frac{f_w}{\Delta \varepsilon} \left[C_{2,w_{(\tau,\varepsilon_h)}} - C_{2,w_{(\tau,\varepsilon_h-1)}} \right] = 0 \tag{21}$$

From the Trogus model,

$$\overline{C}_1 = k_1 C_{1,w}$$
$$\overline{C}_2 = k_2 C_{2,w} \tag{22}$$

A final substitution results in the equation below

$$S_w \frac{\partial C_{1,w_{(\tau,\varepsilon_h)}}}{\partial \tau} + 2 \frac{\partial (k_1 C_{1,w})}{\partial \tau} + \frac{f_w}{\Delta \varepsilon} \left[C_{1,w_{(\tau,\varepsilon_h)}} - C_{1,w_{(\tau,\varepsilon_h-1)}} \right] = 0$$

$$S_w \frac{\partial C_{1,w_{(\tau,\varepsilon_h)}}}{\partial \tau} + 2 k_1 \frac{\partial C_{1,w}}{\partial \tau} + \frac{f_w}{\Delta \varepsilon} \left[C_{1,w_{(\tau,\varepsilon_h)}} - C_{1,w_{(\tau,\varepsilon_h-1)}} \right] = 0$$

$$\left(S_w + 2 k_1 \right) \frac{\partial C_{1,w}}{\partial \tau} + \frac{f_w}{\Delta \varepsilon} \left[C_{1,w_{(\tau,\varepsilon_h)}} - C_{1,w_{(\tau,\varepsilon_h-1)}} \right] = 0$$

and

$$S_w \frac{\partial C_{2,w_{(\tau,\varepsilon_h)}}}{\partial \tau} + 2 \frac{\partial (k_2 C_{2,w})}{\partial \tau} + \frac{f_w}{\Delta \varepsilon} \left[C_{2,w_{(\tau,\varepsilon_h)}} - C_{2,w_{(\tau,\varepsilon_h-1)}} \right] = 0$$

$$\left(S_w + 2 k_2 \right) \frac{\partial C_{2,w}}{\partial \tau} + \frac{f_w}{\Delta \varepsilon} \left[C_{2,w_{(\tau,\varepsilon_h)}} - C_{2,w_{(\tau,\varepsilon_h-1)}} \right] = 0 \tag{23}$$

Application of Orthogonal Collocation to the Solution of Model Equations

Equation (9) can be written as

$$S_w \frac{\partial C'_{i,w}}{\partial \tau} + 2 \frac{\partial \overline{C'_i}}{\partial \tau} + f_w(\tau, \varepsilon_h) \times \left\{ \frac{C'_{i,w}(\tau, \varepsilon_h) - C'_{i,w}(\tau, \varepsilon_{h-1})}{\Delta \varepsilon} \right\} = 0$$
(24)

$$S_w \frac{\partial [\phi C_{i,w}]}{\partial \tau} + 2 \frac{\partial [\rho(1-\phi)\overline{C_i}]}{\partial \tau} + f_w(\tau, \varepsilon_h) \times \left\{ \frac{[\phi C_{i,w}](\tau, \varepsilon_h) - [\phi C_{i,w}](\tau, \varepsilon_{h-1})}{\Delta \varepsilon} \right\} = 0$$
(25)

$$\phi S_w \frac{\partial C_{i,w}}{\partial \tau} + 2\rho(1-\phi) \frac{\partial \overline{C_i}}{\partial \tau} + \phi f_w(\tau, \varepsilon_h) \times \left\{ \frac{C_{i,w}(\tau, \varepsilon_h) - C_{i,w}(\tau, \varepsilon_{h-1})}{\Delta \varepsilon} \right\} = 0$$
(26)

Now, from the Trogus model,

$$\overline{C_i} = \kappa_i C_{i,w}$$
(27)

$$\phi S_w \frac{\partial C_{i,w}}{\partial \tau} + 2\rho(1-\phi) \frac{\partial (\kappa_i C_{i,w})}{\partial \tau} + \phi f_w(\tau, \varepsilon_h) \times \left\{ \frac{C_{i,w}(\tau, \varepsilon_h) - C_{i,w}(\tau, \varepsilon_{h-1})}{\Delta \varepsilon} \right\} = 0$$
(28)

$$\phi S_w \frac{\partial C_{i,w}}{\partial \tau} + 2\kappa_i \rho(1-\phi) \frac{\partial C_{i,w}}{\partial \tau} + \phi f_w(\tau, \varepsilon_h) \times \left\{ \frac{C_{i,w}(\tau, \varepsilon_h) - C_{i,w}(\tau, \varepsilon_{h-1})}{\Delta \varepsilon} \right\} = 0$$
(29)

$$\phi S_w \frac{\partial C_{i,w}}{\partial \tau} + 2\kappa_i \rho(1-\phi) \frac{\partial C_{i,w}}{\partial \tau} + \phi f_w(\tau, \varepsilon_h) \frac{\partial C_{i,w}}{\partial \varepsilon} = 0$$
(30)

$$\left[\phi S_w + 2\kappa_i \rho(1-\phi) \right] \frac{\partial C_{i,w}}{\partial \tau} + \phi f_w(\tau, \varepsilon_h) \frac{\partial C_{i,w}}{\partial \varepsilon} = 0$$
(31)

Let

$$R = \left[\phi S_w + 2\kappa_i \rho(1-\phi) \right]$$

$$B = \phi f_w$$

The above equations now become:

$$R \frac{\partial C}{\partial \tau} + B \frac{\partial C}{\partial \varepsilon} = 0$$
(32)

where C is a function of both e (dimensionless distance) and (dimensionless time).

Using the method of orthogonal collocation, let C be approximated by the expression

$$C(\tau, \varepsilon) = \sum_{I=1}^{N+1} C_I(\tau) X_J(\varepsilon_I) \tag{33}$$

Equation (33) can now be expressed as follows:

$$R\frac{\partial C}{\partial \tau} + B\frac{\partial}{\partial \varepsilon} \sum_{I=1}^{N+1} C_I(\tau) X_J(\varepsilon_I) = 0 \tag{34}$$

$$R\frac{\partial C}{\partial \tau} + B\sum_{I=1}^{N+1} \frac{\partial}{\partial \varepsilon}\left[C_I(\tau) X_J(\varepsilon_I) \right] = 0 \tag{35}$$

$$R\frac{\partial C}{\partial \tau} + B\sum_{I=1}^{N+1} \frac{\partial}{\partial \varepsilon}\left[X_J(\varepsilon_I) \right] \cdot C_I(\tau) = 0 \tag{36}$$

$$a_{JI} = \frac{\partial}{\partial \varepsilon} X_J(\varepsilon_I) \tag{37}$$

$$R\frac{\partial C_J}{\partial \tau} + B\sum_{I=1}^{N+1} a_{JI} C_I = 0 \tag{38}$$

$$\frac{\partial C_J}{\partial \tau} + \frac{B}{R}\sum_{I=1}^{N+1} a_{JI} C_I = 0 \tag{39}$$

$$\frac{\partial C_J}{\partial \tau} = -\frac{B}{R}\sum_{I=1}^{N+1} a_{JI} C_I \tag{40}$$

For $I = 1, 2, 3, 4, \cdots, N+1$

Therefore,

$$\frac{\partial C_J}{\partial \tau} = -\frac{B}{R}\left[a_{J1}C_1 + a_{J2}C_2 + a_{J3}C_3 + a_{J4}C_4 + \cdots + a_{JN+1}C_{N+1}\right]$$

(41)

Again $J = 1, 2, 3, 4, \cdots, N+1$

Therefore the following system of ODE's can be generated

$$\frac{\partial C_1}{\partial \tau} = -\frac{B}{R}\left[a_{11}C_1 + a_{12}C_2 + a_{13}C_3 + a_{14}C_4 + \cdots + a_{1N+1}C_{N+1}\right]$$

$$\frac{\partial C_2}{\partial \tau} = -\frac{B}{R}\left[a_{21}C_1 + a_{22}C_2 + a_{23}C_3 + a_{24}C_4 + \cdots + a_{2N+1}C_{N+1}\right]$$

$$\frac{\partial C_3}{\partial \tau} = -\frac{B}{R}\left[a_{31}C_1 + a_{32}C_2 + a_{33}C_3 + a_{34}C_4 + \cdots + a_{3N+1}C_{N+1}\right]$$

$$\frac{\partial C_4}{\partial \tau} = -\frac{B}{R}\left[a_{41}C_1 + a_{42}C_2 + a_{43}C_3 + a_{44}C_4 + \cdots + a_{4N+1}C_{N+1}\right]$$

$$\vdots$$

$$\frac{\partial C_{N+1}}{\partial \tau} = -\frac{B}{R}\left[a_{N+11}C_1 + a_{N+12}C_2 + a_{N+13}C_3 + a_{N+14}C_4 + \cdots + a_{N+1N+1}C_{N+1}\right]$$

(42)

In matrix form, we have the following expression.

$$\begin{bmatrix} \frac{\partial C_1}{\partial \tau} \\ \frac{\partial C_2}{\partial \tau} \\ \frac{\partial C_3}{\partial \tau} \\ \frac{\partial C_4}{\partial \tau} \\ \vdots \\ \frac{\partial C_{N+1}}{\partial \tau} \end{bmatrix} = -\frac{B}{R}\begin{bmatrix} a_{11} & a_{12} & a_{13} & a_{14} & \cdots & a_{1N+1} \\ a_{21} & a_{22} & a_{23} & a_{24} & \cdots & a_{2N+1} \\ a_{31} & & & & & a_{3N+1} \\ a_{41} & & & & & a_{4N+1} \\ \vdots & \vdots & & & & \vdots \\ a_{N+11} & a_{N+12} & \cdots & \cdots & \cdots & a_{N+1N+1} \end{bmatrix}\begin{bmatrix} C_1(\tau) \\ C_2(\tau) \\ C_3(\tau) \\ C_4(\tau) \\ \vdots \\ C_{N+1}(\tau) \end{bmatrix}$$

(43)

Similarly, the following expression defines a_{JI} [29] [30]

$$a_{JI} = \begin{cases} \dfrac{1}{2}\dfrac{P_{N+1}^{(2)}(\varepsilon_I)}{P_{N+1}^{(1)}(\varepsilon_I)} & \text{for } J = I \\[4mm] \dfrac{1}{\varepsilon_I - \varepsilon_J}\dfrac{P_{N+1}^{(1)}(\varepsilon_I)}{P_{N+1}^{(1)}(\varepsilon_J)} & \text{for } I \neq J \end{cases}$$

(44)

Where

$$P_J(\varepsilon) = (\varepsilon - \varepsilon_J) P_{J-1}(\varepsilon); \quad J = 1, 2, 3, \cdots, N+1$$

$$P_J^{(1)}(\varepsilon) = (\varepsilon - \varepsilon_J) P_{J-1}^{(1)}(\varepsilon) + P_{J-1}(\varepsilon)$$

$$P_J^{(2)}((\varepsilon) = (\varepsilon - \varepsilon_J) P_{J-1}^{(2)}(\varepsilon) + 2 P_{J-1}^{(1)}(\varepsilon)$$

$$P_0^{(1)}(\varepsilon) = P_0^{(2)}(\varepsilon) = 0$$

$$P_0(\varepsilon) = 1 \tag{45}$$

Recall that the elements of the matrix can be generated from the following Lagrange polynomial

$$a_{ij} = \frac{dl_j(x_i)}{dx} = \begin{cases} \dfrac{1}{2} \dfrac{P_{N+1}^{(2)}(x_i)}{P_{N+1}^{(1)}(x_i)} & j = i \\[3mm] \dfrac{1}{x_i - x_j} \dfrac{P_{N+1}^{(1)}(x_i)}{P_{N+1}^{(1)}(x_j)} & i \neq j \end{cases} \tag{46}$$

For i = j, the elements here refer to the leading diagonal of the matrix to be generated.

For i ≠ j, the elements here refer to all other elements of the matrix.

Also, the following recurrence relations are defined below

$$p_o(x) = 1$$

$$P_j(x) = (x - x_j) P_{j-1}(x)$$

$$P_j^{(1)}(x) = (x - x_j) P_{j-1}^{(1)}(x) + P_{j-1}(x)$$

$$P_j^{(2)}(x) = (x - x_j) P_{j-1}^{(2)}(x) + 2 P_{j-1}^{(1)}(x) \tag{47}$$

For $j = 2, 3, 4, \cdots, N+1.$

The following substitutions and manipulations will now be made to redefine Equation (46).

Substituting the recurrence relations into Equation (46) yields

$$a_{ij} = \begin{cases} \dfrac{1}{2}\left[\dfrac{\left(x_i - x_j\right)P_{j-1}^{(2)}\left(x_i\right) + 2P_{j-1}^{(1)}\left(x_i\right)}{\left(x_i - x_j\right)P_{j-1}^{(1)}\left(x_i\right) + P_{j-1}\left(x_i\right)}\right] & j = i \\[4ex] \dfrac{1}{x_i - x_j}\left[\dfrac{\left(x_i - x_j\right)P_{j-1}^{(1)}\left(x_i\right) + P_{j-1}\left(x_i\right)}{\left(x_j - x_j\right)P_{j-1}^{(1)}\left(x_j\right) + P_{j-1}\left(x_j\right)}\right] & i \neq j \end{cases}$$

(48)

Now, some terms will be cancelled out.

Since j = i, ($x_i - x_j$) = 0,

And

($x_j - x_j$) = 0

$$a_{ij} = \begin{cases} \dfrac{1}{2}\left[\dfrac{2P_{j-1}^{(1)}\left(x_i\right)}{P_{j-1}\left(x_i\right)}\right] & j = i \\[4ex] \dfrac{1}{x_i - x_j}\left[\dfrac{\left(x_i - x_j\right)P_{j-1}^{(1)}\left(x_i\right) + P_{j-1}\left(x_i\right)}{P_{j-1}\left(x_j\right)}\right] & i \neq j \end{cases}$$

(49)

The above becomes

$$a_{ij} = \begin{cases} \left[\dfrac{P_{j-1}^{(1)}\left(x_i\right)}{P_{j-1}\left(x_i\right)}\right] & j = i \\[4ex] \dfrac{\left(x_i - x_j\right)P_{j-1}^{(1)}\left(x_i\right)}{\left(x_i - x_j\right)P_{j-1}\left(x_j\right)} + \dfrac{1}{x_i - x_j}\left[\dfrac{P_{j-1}\left(x_i\right)}{P_{j-1}\left(x_j\right)}\right] & i \neq j \end{cases}$$

(50)

This becomes

$$a_{ij} = \begin{cases} \left[\dfrac{P_{j-1}^{(1)}\left(x_i\right)}{P_{j-1}\left(x_i\right)}\right] & j = i \\[4ex] \dfrac{P_{j-1}^{(1)}\left(x_i\right)}{P_{j-1}\left(x_j\right)} + \dfrac{1}{x_i - x_j}\left[\dfrac{P_{j-1}\left(x_i\right)}{P_{j-1}\left(x_j\right)}\right] & i \neq j \end{cases}$$

(51)

Rewriting the above in terms of epsilon, ( )

$$a_{ij} = \begin{cases} \left[\dfrac{P_{j-1}^{(1)}(\varepsilon_i)}{P_{j-1}(\varepsilon_i)} \right] & j = i \\[2em] \dfrac{P_{j-1}^{(1)}(\varepsilon_i)}{P_{j-1}(\varepsilon_j)} + \dfrac{1}{\varepsilon_i - \varepsilon_j} \left[\dfrac{P_{j-1}(\varepsilon_i)}{P_{j-1}(\varepsilon_j)} \right] & i \neq j \end{cases}$$

(52)

The matrix now looks like this

$$a_{11} = \frac{P_0^{(1)}(\varepsilon_1)}{P_0(\varepsilon_1)}$$

$$a_{12} = \frac{P_1^{(1)}(\varepsilon_1)}{P_1(\varepsilon_2)} + \frac{1}{\varepsilon_1 - \varepsilon_2} \frac{P_1(\varepsilon_1)}{P_1(\varepsilon_2)}$$

$$a_{13} = \frac{P_2^{(1)}(\varepsilon_1)}{P_2(\varepsilon_3)} + \frac{1}{\varepsilon_1 - \varepsilon_2} \frac{P_2(\varepsilon_1)}{P_2(\varepsilon_3)}$$

$$a_{21} = \frac{P_0^{(1)}(\varepsilon_2)}{P_0(\varepsilon_1)} + \frac{1}{\varepsilon_2 - \varepsilon_1} \frac{P_0(\varepsilon_2)}{P_0(\varepsilon_1)}$$

$$a_{22} = \frac{P_1^{(1)}(\varepsilon_2)}{P_1(\varepsilon_1)}$$

$$a_{23} = \frac{P_2^{(1)}(\varepsilon_2)}{P_2(\varepsilon_3)} + \frac{1}{\varepsilon_2 - \varepsilon_3} \frac{P_2(\varepsilon_2)}{P_2(\varepsilon_3)}$$

$$a_{31} = \frac{P_0^{(1)}(\varepsilon_3)}{P_0(\varepsilon_1)} + \frac{1}{\varepsilon_3 - \varepsilon_1} \frac{P_0(\varepsilon_3)}{P_0(\varepsilon_1)}$$

$$a_{32} = \frac{P_1^{(1)}(\varepsilon_3)}{P_1(\varepsilon_2)} + \frac{1}{\varepsilon_3 - \varepsilon_2} \frac{P_1(\varepsilon_3)}{P_1(\varepsilon_2)}$$

$$a_{32} = \frac{P_2^{(1)}(\varepsilon_3)}{P_2(\varepsilon_3)}$$

(53)

The recurrence relations below will again be used to evaluate the terms of the matrix

$$p_o(\varepsilon) = 1$$

$$P_j(\varepsilon) = (\varepsilon - \varepsilon_j) P_{j-1}(\varepsilon)$$

$$P_j^{(1)}(\varepsilon) = (\varepsilon - \varepsilon_j) P_{j-1}^{(1)}(\varepsilon) + P_{j-1}(\varepsilon)$$

$$P_0^{(1)}(\varepsilon) = 0$$

(54)

Let e assume the range e = [0:0.01:0.09]

Where

$$\varepsilon_1 = 0$$

(55)

$$\varepsilon_2 = 0.01$$

(56)

$$\varepsilon_3 = 0.02$$

(57)

RESULTS

The reservoir response, as predicted by the simulation on the basis of the theory of coherence, is compared with the numerical predictions obtained using traditional finite difference method and orthogonal collocation. For the case studies, data from a hypothetical well and an existing Nigerian well data were used for the validation of the simulation design. The main objective of these case studies has been to demonstrate that the mathematical techniques of orthogonal collocation, finite difference and coherent theory in the context of simulator design can be used to obtain wave behaviour in a reservoir. A gradually increasing level of complexity is introduced, representing a range of systems from aqueous phase flow, to surfactant chromatography in two phase flow and to surfactant chromatography in two dimensional porous medium. The initial and injected surfactant compositions corresponding to Cases 1, 2 and 3 are shown in Table 5. The rock and fluid properties are listed in Tables 1-4. These properties are assumed uniform for convenience.

The two fluid phases consisted of a water phase and an oil phase, which, for convenience are considered incompressible. The density of oil, the viscosity of oil, the salinity of water, and the formation volume factor of oil and water are listed in Table 2. All cases mentioned above were run by using anionic sodium dodecyl sulfate (SDS) and cationic dodecyl pyridinium chloride (DPC) as surfactants.

The system of equations is complete with the equations representing physical properties of the fluids and the rock. Physical properties described here are: 1) phase behaviour; 2) interfacial tension between fluid phases; 3) residual phase saturations; 4) relative permeabilities; 5) rock wettabiliy; 6) phase viscosities; 7) capillary pressure; 8) adsorption and 9) dispersion. From a physico-chemical point of view, there are three components—wa- ter, petroleum and chemical. They are in fact, pseudo-components, since each one consists of several pure components. Petroleum is a complex mixture of many hydrocarbons. Water is actually brine and contains dissolved salts. Finally, the chemical contains different kinds of surfactants.

These three pseudo-components are distributed between two phases—the oleic phase and the aqueous phase. The chemical has an amphiphilic character. It makes the oleic phase at least partially miscible with water or the aqueous phase partially miscible with petroleum.

Interfacial tension depends on the phase behaviour as the surfactant partitions between the two phases. Residual phase saturation decreases as interfacial tension decreases. Relative permeability parameters depend on residual phase saturations. Phase viscosities are functions of the volume fraction of the components in each fluid phase. Therefore, the success or failure of surfactant flooding processes depends on phase behaviour. Also, phase behaviour influences all the other physical properties, and each of them, in turn influences oil recovery.

For a two-phase flow of water and oil, where no surfactant partitions into the oleic phase, the same scenario is obtained as the one dimensional injection for Cases 1 and 2. The bed has an initial water saturation of 0.3, and is flooded with an aqueous surfactant solution. The numerical profiles agree with the coherent wave profiles. The effect of the two-phase flow is to elongate the waves, leading to a larger region of constant state and earlier breakthrough of the fast wave.

Table 5: Conditions for case studies of surfactant chromatography Hankins and Harwell [25]

Case	Injected composition: CC1 (mol/ m³bed)	Injected composition: CC2 (mol/ m³bed)	Initial composition: C1 (mol/ m³bed)	Initial composition: C2 (mol/ m³bed)
1	0.17	0.013	0.21	0.181
2	0.042	0.115	0	0
3	0.66	0.875	0.35	0.15

Figure 1(a) is the result obtained for solving Equation (4) using the three methods finite difference, orthogonal collocation, coherent theory. The graph is for the composition profile for one dimensional two-phase chromatography initially equilibrated with a composition $C_1 = 0.21$, $C_2 = 0.181$ and is then injected with a composition $C_1 = 0.17$, $C_2 = 0.013$ (Riemann type problem Case 1, (refer toTable 5)). In Figure 1(a), the profile C_1 of finite difference shows a steady rise from $C_1 = 0.17$ to $C_1 = 0.21$ and then remain constant at this concentration.

The profile C_1 of the orthogonal collocation increases steadily from $C_1 = 0.17$ to $C_1 = 0.21$ at distance 0.1 epsilon maintaining a constant state to distance 0.3 epsilon. After this, it started declining from $C_1 = 0.21$ to $C_1 = 0.07$ at distance 0.5 epsilon before rising back to attain a constant state with the finite difference. The profile C_1 of the coherent theory on the other hand started with a constant state, then declined before it reaches a constant state and then rises again to attain another constant state with the other profiles. Similarly, the profile C_2 of finite difference increased steadily from $C_2 = 0.017$ to a constant state of $C_2 = 0.18$. The orthogonal collocation for C_2 starts at $C_2 = 0.01$ for a short constant state and then rises steadily to $C_2 = 0.18$ to attain another short constant state from 0.2 to 0.3 epsilon. From here it depressed to $C_2 = 0.07$ before rising back to $C_2 = 0.18$ and then attain a constant state with finite difference. The profile C_2 of coherent theory starts with a short constant state, then in- creases readily to $C_2 = 0.05$ for another constant state from where it rises up to the final region of constant state with the other profiles.

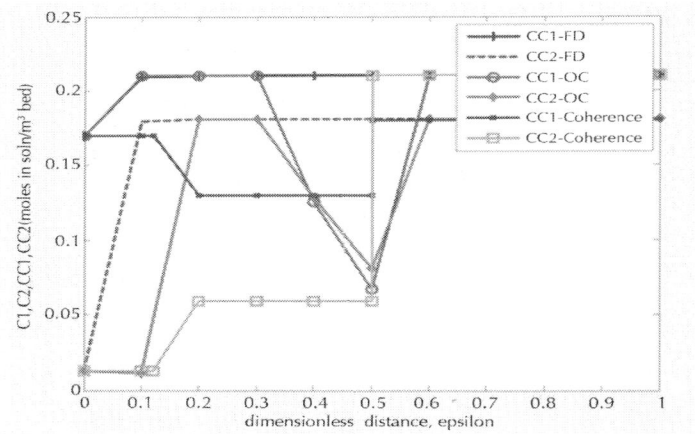

(a)

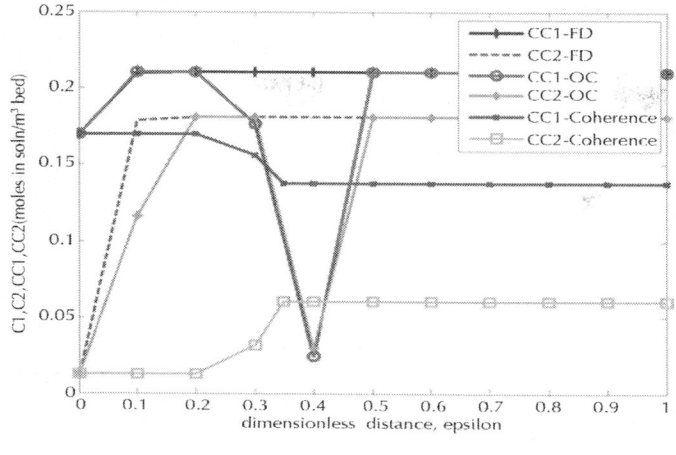

(b)

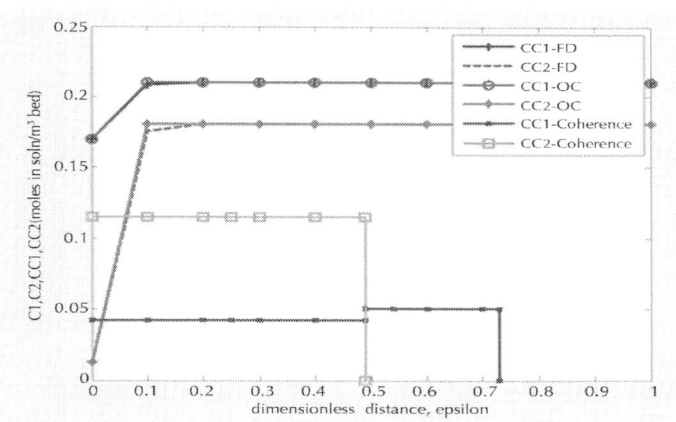

(c)

Figure 1: (a) Case 1. C_1, C_2 vs epsilon at   = 0.5. Bed composition profile for one-dimensional two-phase chromatography; Case 1, at one-half pore volume injected. The plots are for three methods: Orthogonal collocation (OC), finite difference (FD) and coherent theory (CT); (b) Case 1 C_1, C_2 vs epsilon at   = 1.0. Bed composition profile for one-dimensional two-phase chromatography; Case 1, at one pore volume injected. The plots are for three methods: Orthogonal collocation (OC), finite difference (FD) and coherent theory (CT). (c) Case 1 C_1, C_2 vs epsilon at $= 2.0$. Bed composition profile for one-dimensional two-phase chromatography; Case 1, at two pore volumes injected. The plots are for three methods: Orthogonal collocation (OC), finite difference (FD) and coherent theory (CT).

Figure 1(b) shows the result obtained for solving Equation (4) by using orthogonal collocation and finite difference as the numerical technique. The graph is for the bed composition profile for one dimensional two-phase chromatography for Case 1 at one pore volume injected. In this case also, the adsorbing porous medium is initially equibrated with a composition $C_1 = 0.21$, $C_2 = 0.181$ (concentrations normalized as moles in solution per m^3 of bed) and is then injected with a composition $C_1 = 0.17$, $C_2 = 0.013$ (Riemann-type problem: Case 1 (refer to Table 5). The profile C_1 for finite difference indicates a rise in concentration from $C_1 = 0.17$ to 0.21 after which the concentration maintained a constant state. The profile of C_1 for the orthogonal collocation also rise from $C_1 = 0.17$ to $C_1 = 0.21$ but falls to 0.03 epsilon at distance 0.4 epsilon and then increased thereafter to constant state as maintained in the C_1 profile for finite difference. The profile C_1 of coherent theory started with a constant concentration and then decreased gradually to attain another region of constant state. The profile C_2 of finite difference increased steadily from $C_2 = 0.02$ to attain constant state at 0.18 epsilon. Also the profile of C_2 of the orthogonal collocation increases gradually from $C_2 = 0.02$ to $C_1 = 0.18$ at distance 0.2 for a short constant state and then decline to a low value of C_2 $= 0.02$ at distance 0.4 epsilon before rising back to reach a constant state with the finite difference profile. The profile C_2 of coherent theory started with a constant state and gradually rises to a constant state

The bed composition profile for one dimensional two-phase chromatography for Case 1 at two pore volume injected is shown in Figure 1(c). This is the result obtained for solving Equation (4) by using these three techniques; finite difference, orthogonal collocation and coherent theory. The adsorbing porous medium is initially equibrated with a composition $C_1 = 0.21$, $C_2 = 0.181$ (concentrations normalized as moles in solution per m^3 of bed) and is then injected with a composition $C_1 = 0.17$, $C_2 = 0.013$ (Riemann-type problem: Case 1 (refer to Table

5)). The profile C_1 of finite difference and the profile C_1 of orthogonal collocation indicate that there is steady increase from $C_1 = 0.17$ to $C_1 = 0.21$ at distance 0.1 epsilon and then attained a constant state for both profiles. However, the profile C_1 of coherent theory shows a constant state of concentration at $C_1 = 0.04$ before having a self sharpening shock at $C_1 = 0.05$, and eigenvalue $= 0.499$. It then continues at this constant state and then decreased to zero at distance 0.73. Similarly, the profile C_2 of finite difference shows a steady rise from $C_2 = 0.02$ to $C_2 = 0.18$ and then maintained a constant state. Also, the profile C_2 for orthogonal collocation, follows the same pattern, which indicates an increase from $C_2 = 0.02$ to $C_2 = 0.18$ and then attained a constant state. The orthogonal collocation profiles match the finite difference profiles. However, the profile C_2 of coherent theory shows a constant state of concentration but has no self sharpening at $C_2 = 0.00$ in this case before its termination at distance 0.499 epsilon.

Figure 2(a) shows the bed concentration profiles for one dimensional two-phase chromatography for Case 2 at one pore volume injected in the porous medium initially devoid of surfactant and then injected with a mixture C_1

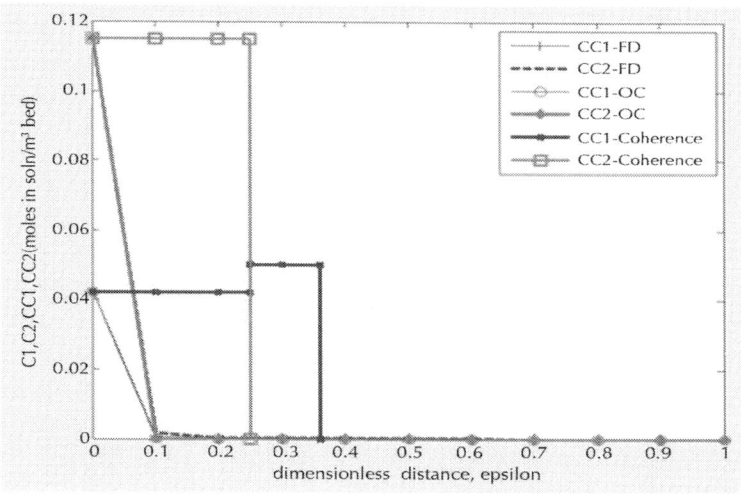

(a)

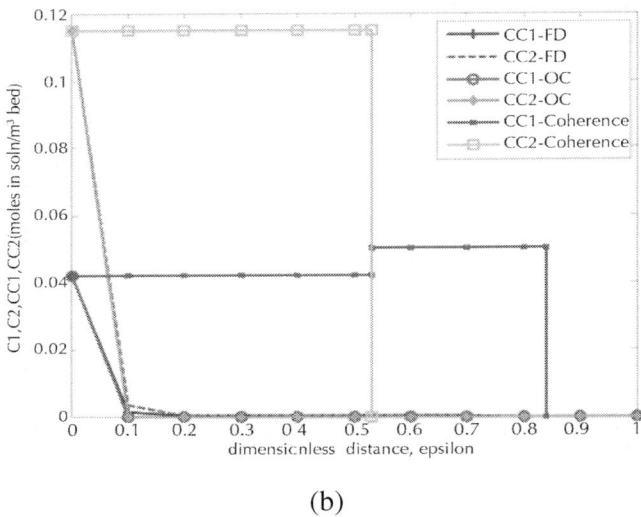

(b)

Figure 2: (a) Case 2, C_1, C_2 vs epsilon at = 1.0. Bed composition profile for one-dimensional two-phase chromatography; Case 2, at one pore volume injected. The plots are for three methods: Orthogonal collocation (OC), finite difference (FD) and coherent theory (CT); (b) Case 2. C_1, C_2 vs epsilon at = 2.0. Bed composition profile for one-dimensional two-phase chromatography; Case 2, at two pore volumes injected. The plots are for three methods: Orthogonal collocation (OC), finite difference (FD) and coherent theory (CT).

= 0.042, C_2 = 0.115 (Riemann-type problem, Case 2 (refer to Table 5), with the numerical result obtained for solving Equation (4) by using three different techniques; orthogonal collocation, finite difference and coherent theory. The profile C of finite difference shows steady decline from C_1 = 0.04 to a constant state, the same as for orthogonal collocation. However, the profile C_1 of coherent theory shows a constant state of concentration at C_1 = 0.04 before having a self sharpening shock at C_1 = 0.048 and eigenvalue = 0.26. It then continues with a constant state and decreased to zero at distance 0.37. The profile C_2 of finite difference decreased steadily from C_2 = 0.119 to C_2 = 0.001 and then to a constant state as for C_1; a. similar behaviour was observed for orthogonal collocation. Again, the profile C_2 was different for coherent theory where a constant state was initially observed with no self sharpening at C_2 = 0.00.

The bed composition profile for one dimensional two-phase chromatography for Case 2 at two pore volume injected is shown in Figure 2(b). This is the result obtained for solving Equation (4) by using the three techniques; finite difference, orthogonal collocation and coherent theory. The profile of C_1 for finite difference shows steady decline from $C_1 = 0.04$ to a constant state; similar behaviour was observed for orthogonal collocation. A difference was again observed for the profile C_1 of coherent theory where the profile shows a constant state of concentration at $C_1 = 0.04$ before having a self sharpening shock of $C_1 = 0.046$, and eigenvalue $= 0.53$. It then continues with a constant state and decreased to zero at distance 0.84 epsilon. The profile C_2 of finite difference decreased steadily from $C_2 = 0.119$ to $C_2 = 0.001$ and then to a constant state as for C_1; a similar behaviour was observed for orthogonal collocation.

In Figure 2(b), the profiles C_1 of orthogonal collocation, finite difference and coherent theory follow the same pattern as that in Figure 2(a). Similarly, the profiles C_2 of finite difference, orthogonal collocation, coherent theory in Figure 2(a) have the same pattern as those in Figure 2(b).

There are two regions of moderate change corresponding to two "fronts". These are the leading edge of the surfactant and the solubilization (or miscible) front as the concentrations jump to their injected values. Physically, this region corresponds to the very rapid increase in the relative permeability of the aqueous phase due to the decrease in interfacial tension. This is, of course, what the surfactant is designed to do, and is a physically desirable feature of the process.

Oil Recovery

It is necessary to generate averaged relative permeability curves which are functions of the thickness averaged water saturation and use these in the oil recovery calculations. At a constant temperature, the oil and water viscosities have fixed values and are strictly functions of the water saturation as related through the relative permeabilities.

Table 6 shows the values of water saturation and viscosities for Case A (hypothetical reservoir data) and Case B (Nigerian reservoir data) from which the relative permeability curves for cases A and B are plotted respectively (see Figure 3 and Figure 4).

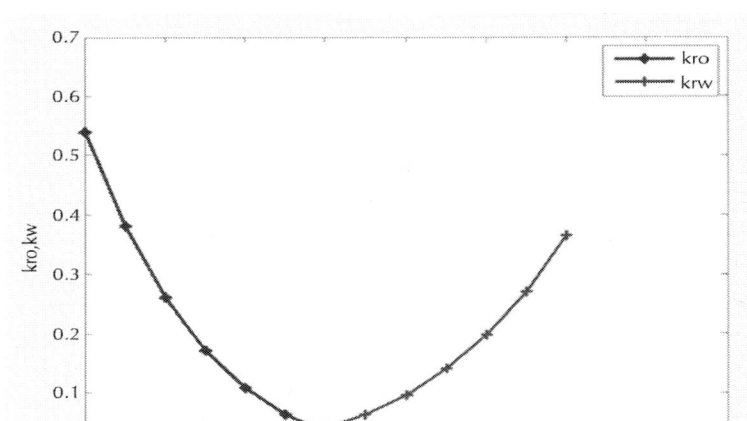

Figure 3: Graph of k_{ro} and k_{rw} against S_w for Case A. Effective and corresponding relative permeability as functions of water saturation for Case A.

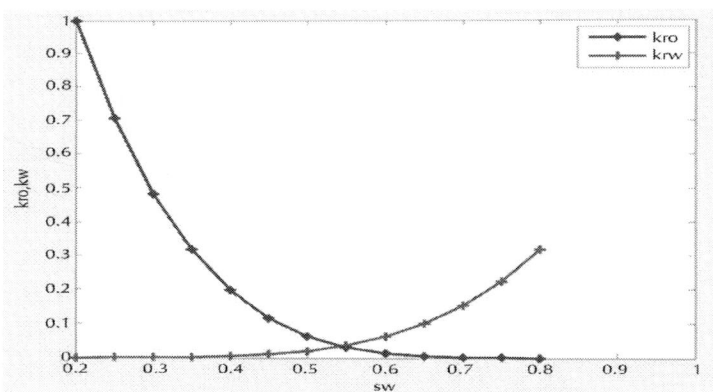

Figure 4: Graph of k_{ro} and k_{rw} against S_w for Case B. Effective and corresponding relative permeability as functions of water saturation for Case B.

Table 6: Water saturations and viscosities for Cases A and B.

S_w	μ_o		μ_w		S_{ro}
	Case A	Case B	Case A	Case B	
0.2	5	0.4	1	0.3	0.2
0.25	5	0.4	1	0.3	0.2
0.3	5	0.4	1	0.3	0.2
0.35	5	0.4	1	0.3	0.2
0.4	5	0.4	1	0.3	0.2
0.45	5	0.4	1	0.3	0.2
0.5	5	0.4	1	0.3	0.2
0.55	5	0.4	1	0.3	0.2
0.6	5	0.4	1	0.3	0.2
0.65	5	0.4	1	0.3	0.2
0.7	5	0.4	1	0.3	0.2
0.75	5	0.4	1	0.3	0.2
0.8	5	0.4	1	0.3	0.2

Table 7 shows the values of fractional flow and water saturation in the reservoir for Case A and Case B. These values are used to obtain fractional flow plots for both Cases A and B as shown in Figure 5.

Figure 3 shows the saturation dependence of the effective permeabilities of oil and water. The plot shows both permeabilities as functions of the water saturation alone since the oil saturation is related to the water saturation by a simple relationship $S_o = 1 - S_w$. On the effective permeability of water (k_{rw}) curve, $S_w = S_{wc} = 0.2$ on the S_w axis. At $S_w = 1$, $k_w = 1.0$. Similarly, for the effective permeability for oil curve $S_w = S_{wc} = 0.2$ At $S_w = 1$, $k_{ro} = 0.55$. The curves are used to describe the displacement of oil by water taking into consideration the manner in which the fluid saturations are distributed with respect to thickness as they simultaneously move through the reservoir. Surfactants dissolved in minute quantities of water have significant effect on displacement of oil and increase the oil saturation above Sor. This result in moving from point S_{or}, $K_{or} = 1$ on the normal relative permeability curve thereby raising the saturation above the residual level to render the oil mobile.

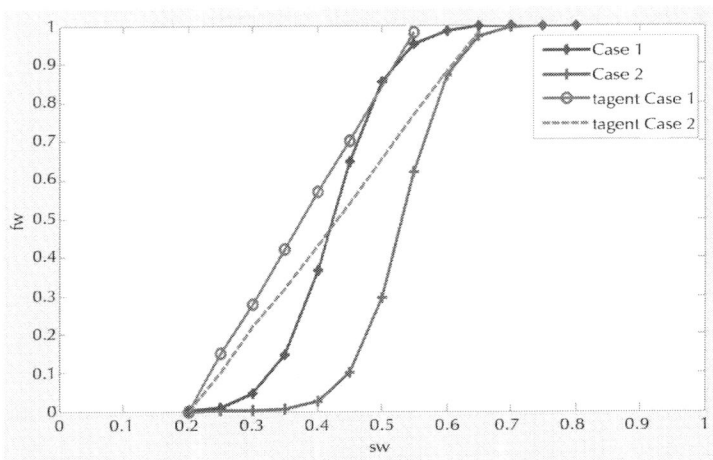

Figure 5. Fractional flow plots for oil-water viscosity ratios. Graph of fw against sw.

Table 7: Fractional flow and water saturation for cases A and B.

F_w		K_{ro}		K_{rw}		S_{iw}		S_w
Case A	Case B	Case A	Case B	Case A	Case B	Case A	Case B	
0.0014	0	0.5398	1	0.0002	0	0.1	0.2	0.2
0.01	0	0.3811	0.7061	0.0008	0	0.1	0.2	0.25
0.0447	0.0007	0.2603	0.4823	0.0024	0.0002	0.1	0.2	0.3
0.1484	0.0052	0.1708	0.3164	0.006	0.0012	0.1	0.2	0.35
0.3667	0.0257	0.1066	0.1975	0.0123	0.0039	0.1	0.2	0.4
0.6466	0.099	0.0625	0.1158	0.0229	0.0095	0.1	0.2	0.45
0.8526	0.2967	0.0337	0.0625	0.039	0.0198	0.1	0.2	0.5
0.9505	0.6184	0.0163	0.0301	0.0625	0.0366	0.1	0.2	0.55
0.9862	0.871	0.0067	0.0123	0.0953	0.0625	0.1	0.2	0.6
0.997	0.9716	0.0021	0.0039	0.1395	0.1001	0.1	0.2	0.65
0.9996	0.9962	0.0004	0.0008	0.1975	0.1526	0.1	0.2	0.7
1	0.9998	0	0	0.2721	0.2234	0.1	0.2	0.75
1	1	0	0	0.366	0.3164	0.1	0.2	0.8

Figure 4 shows the saturation dependence of the effective permeabilities of oil and water. The plot shows both permeabilities as functions of the water saturation alone since the oil saturation is related to the water saturation by a simple relationship $S_o = 1 - S_w$. On the effective permeability of water (k_{rw}) curve $S_w = S_{wc} = 0.2$ on the S_w axis. At $S_w = 1$ $k_w = 0.7$. Similarly, for the effective permeability for oil curve $S_w = S_{wc} = 0.2$ at $S_w = 1$ $k_{ro} = 1.0$ Also, the curves are used to describe the displacement of oil by water taking into consideration the manner in which the fluid saturations are distributed with respect to thickness as they simultaneously move through the reservoir. Surfactants dissolved in minute quantities of water have significant effect on displacement of oil and increase the oil saturation above S_{or}. This result in moving from point S_{or}, $K_{or} = 1$ on the normal relative permeability curve thereby raising the saturation above the residual level to render the oil mobile.

If we compare Figure 3 and Figure 4 for the hypothetical data and the Nigerian oil data, we notice the residual oil saturation curve of Figure 4 being shifted to the right of the residual oil saturation curve of Figure 3. So, when the surfactant contacts the residual oil, the surfactant dissolves in it and then raises the saturation above the residual level. Also, there is a significant reduction in the surface tension between these enlarged oil droplets and the displacing water. Moreover, since the relative permeability of the oil increases from zero upwards, then the residual oil becomes more mobile for the Nigerian oil (Figure 4) than that in Figure 3 (the hypothetical oil). This results in more oil being recovered from Figure 4 than in Figure 3.

Fractional flow values for A and B obtained with the application of Welge's technique at breakthrough are listed in Table 7. Also, the fractional flow curve drawn using the averaged relative permeability curves and the Buckley Leverett/Welge technique to calculate oil recovery is shown inFigure 5. This figure shows the fractional flow plots for different oil water viscosity ratio for cases A and B. The fractional flow plot exhibit the characteristic S shape with saturation limits, S_{wc} and $1 - S_{or}$, between which the fractional flow increases from zero to unity. The tangent to the fractional flow curve from the point $S_w = S_w$; $f_w = 0$ has a point of tangency with co-ordinates $S_w = S_{wbt} = 0.50$; $f_w = f_w|_{swf} = f_{wbt} = 0.86$ for Case A and $S_w = S_{wbt} = 0.61$; $f_w = f_w|_{swf} = f_{wbt} = 0.91$ for Case B (refer to Table 8). The extrapolated tangent intercept the line $f_w = 1$ at the point $f_w = 1$; $S_w = S_{wbt} = 0.55$ for Case A and at point $f_w = 1$; $S_w = S_{wbt} = 0.65$ for Case B. (refer to Table 8). Table 9shows values of shock front, end point relative permeabilities and mobilities ratio obtained fromFigure 5 by applying Welge's technique.

Figure 6 shows the oil recovery in a resevoir as a function of dimensionless water injected for both casE A (hypothetical resevoir) and Case B (Nigerian reservoir). The graph is plotted from the caiculated values of f_{we}, W_{id} and N_{pd} shown in Table 10 and Table 11 The graph shows that the oil recovery curve for Case B rises above that of Case A even with less injected pore volume. This show that the break through occurred much earlier for B. The oil recovery for Case A reservoir is 0.61 while that for Case B reservoir is 0.73.

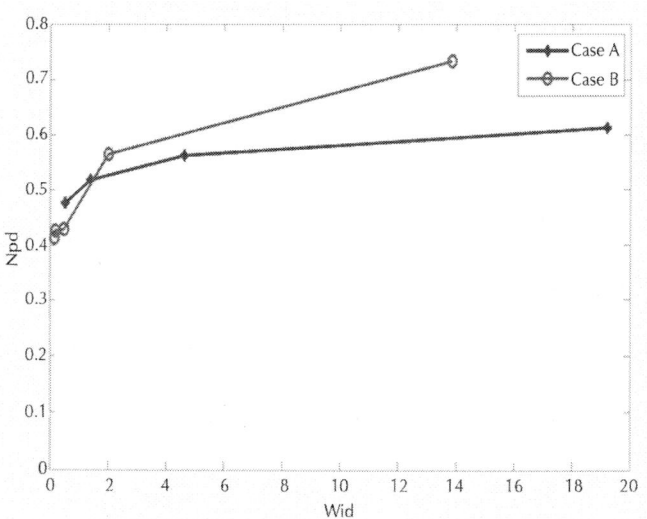

Figure 6: Dimensionless oil recovery as a function of dimensionless injected pore volume. Comparison of the oil recoveries obtained for Case A and Case B reservoirs.

Table 8: Flood front saturation values

Case NO	$S_{w_{bt}}$	$f_{w_{bt}}$	$f_{ws_{bt}}$	$\bar{\bar{S}}_{w_{bt}}$	$N_{pd_{bt}}$ **(PV)**
A	0.50	0.86	0.89	0.55	0.35
B	0.61	0.91	0.94	0.65	0.45

Table 9: Values of shock and end point relative permeabilities

Case NO	μ_o	S_{wf}	$k_{wf}(S_{wf})$	$k_{ro}(S_{wf})$	M_s	M
A	5.0	0.50	0.02	0.015	0.23	2.70
B	1.33	0.61	0.0366	0.0122	0.20	0.90

Table 10: Cumulative oil recovery as function of wid for Case A

S/No	S_{we}	Few	ΔS_{we}	Δf_{we}	$\Delta f_{we}/\Delta S_{we}$	S_{we}	W_{id}
1	0.50	0.8526	-	-	-	-	-
2	0.55	0.9505	0.05	0.0979	1.958	0.525	0.5107
3	0.60	0.9862	0.05	0.0357	0.714	0.575	1.401
4	0.65	0.9970	0.05	0.0108	0.216	0.625	4.630
5	0.70	0.9996	0.05	0.0026	0.052	0.675	19.231
6	0.75	1.0000	0.05	0.0004	0.008	0.725	125.00
7	0.80	1.0000	0.05	0.0000		0.775	

	S_{we}	$S_{we} - S_{wc}$	few	1 − f	W_{id}	N_{pd}
1	-	-	-	-	-	-
2	0.525	0.425	0.9200	0.0800	0.5107	0.4753
3	0.575	0.475	0.9700	0.0300	1.4010	0.5193
4	0.625	0.525	0.9900	0.0100	4.6300	0.5638
5	0.675	0.575	0.9980	0.0020	19.2310	0.6135
6	0.725	0.625	0.9998	0.0002	125.000	0.6500
7	0.775	0.665	1.0000			

Table 11: Cumulative oil recovery as function of wid for Case B

	S_{we}	f_{we}	ΔS_{we}	Δf_{we}	$\Delta f_{we}/\Delta S_{we}$	S_{we}	W_{id}
1	0.5	0.2967	-	-	-	-	-

2	0.55	0.6184	0.05	0.3217	6.434	0.525	0.1554
3	0.60	0.8710	0.05	0.2526	5.052	0.575	0.1979
4	0.65	0.9716	0.05	0.1006	2.012	0.625	0.4970
5	0.70	0.9962	0.05	0.0246	0.492	0.675	2.0330
6	0.75	0.9998	0.05	0.0036	0.072	0.725	13.8800
7	0.80	1.0000	0.05	0.0002	0.004	0.775	250.000

	S_{we}	$S_{we} - S_{wc}$	few	$1 - f$	W_{id}	N_{pd}
1	-	-	-	-	-	-
2	0.525	0.325	0.4400	0.5600	0.1554	0.4120
3	0.575	0.375	0.7450	0.2550	0.1979	0.4255
4	0.625	0.425	0.86500	0.1350	0.4970	0.4921
5	0.675	0.475	0.9550	0.0450	2.0330	0.5665
6	0.725	0.525	0.9850	0.0150	13.880	0.7332
7	0.775	0.575	0.9980	0.0020	250.00	

The tangent to the fractional flow curve from the point $S_w = S_{wc}$, $f_w = 0$ gives the coordinates to obtain parameters to calculate the oil recovery for both Cases A (Hypothetical data) and B (Nigerian oil data).

The values in Table 5 and Table 6 were used to generate the plots of dimensionless oil recovery as a function of injected pore volume in Figure 6. This shows the oil recoveries for both Case A and B reservoirs.

In Figure 6, plot of the oil recovery of reservoir A is compared with that of plot of oil recovery for reservoir B. The comparison shows that, the breakthrough occurs much earlier for Case B, the ultimate recovery is obtained for a much smaller throughput of injected pore volume and sooner than for Case A. The oil recovery for Case A reservoir is 0.61 while that for Case B reservoir is 0.73

DISCUSSION OF RESULTS

The ultimate objectives of the simulator designed in this work are two-fold:

- The prediction of the appropriate surfactant concentrtion necessary for the required enhanced oil recovery from reservoirs, and

- The volume of recovered oil from the reservoir using the hypothetical reservoir oil data and Nigerian oil reservior data.

With regard to item 1 above, the basic physical principle employed by the simulator is that of mass conservation. Usually those quantities are conserved at stock tank conditions and related to reservoir fluid quantities through the pressure dependent parameters. The profiles of three cases (1, 2, 3) in one dimensional aqueous phase chromatography and two-phase chromatography for one, one-half, and two pore volume injected were developed using simulated solutions to model equations. These equations were solved by finite difference (FD), orthogonal collocation (OC) and coherent theory (CT).The use of these methods permit the determination of the relative efficiency of the methods and how well they predict the complex characteristics of the enhanced oil recovery process. We will now discuss the significant results of this work regard to items 1 and 2 above.

We did find out that:

- Injecting a mixture of low concentration aqueous surfactant composition into adsorbing porous medium that is initially injected with high concentration aqueous surfactant composition. This variation may exist in the initial profile or be generated by injection. The initial fluid or previously injected fluid has the composition downstream of the change in amount while the newly injected fluid has the composition upstream of the original variation. The composition route along the bed follows the slow path from the injected composition and then switches to the fast path which leads to the previously injected compòsition. The route passes along paths and follows the paths in the sequence of increasing wave velocities.

- Injecting a mixture of an aqueous composition into a porous medium, initially devoid of surfactant, the expected composition is a self sharpening shock wave. The steepness in all the profiles generated by finite difference (FD), orthogonal collocation (OC), and coherent theory (CT) confirms the self sharpening behaviour. It may be noted in all cases of these nature the waves trajectories gradually fall, as a result of a gradual increase in the associated eigenvalues of the waves as salinity increases. The finite difference (FD) and orthogonal collocation (OC) response essentially agrees with the coherent method except for a slight curvature. This is because the finite difference (FD) and orthogonal collocation (OC) are no longer a good approximation to the shock The consequence of this steepening is that the flows are sharpening, so that they break through both earlier and over a smaller injected volume. For the dependent variables such as component concentration, common velocity exists at each point in the wave, and the associated composition route remains unchanged and the same during relative shifts of waves associated with other dependent variable waves as shown in all the methods. This is in agreement with Helferich [4] .

Injecting a mixture of high concentration of surfactant into adsorbing porous medium that is initially injected with low concentration aqueous surfactant composition yield two types of path. The slow and fast paths. The slow path eigenvalues are close to the fast path with eigenvalues of 1 and the effect of dispersion results in the merging of the two waves. This is due to their spatial positions, and loss of intermediate region of constant state. This region later reappear with less dispersion.

- For a system of two-phase (aqueous and oleic) flow of water and oil in which there is no surfactant partitions into the oleic phase. The effect of two phase flow elongated the wave, leading to a larger region of constant state and earlier breakthrough of fast wave. The shock and wave composition route are identical to that of aqueous phase and has a unique composition path grid which is independent of the saturation of the flowing phases. This is similar to system of a two-phase (aqueous and oleic) flow in which two water soluble components partition between the aqueous and a solid adsorbent phase [20] .

- When surfactant partitions into the oleic phase, the effect is that the adsorption of the surfactant in the solid phase results in the formation of micelles which then break away from the solid phase and move much faster than the surfactant in the oleic phase. The domination of the micelle formation in the aqueous phase is the general effect for both slow and fast phases. The presence of the two partitioning components does not alter the fractional flow relations $f_j(S_j)$ for the two phases. The sum of the aqueous and oleic phase saturations must add up to one wave of the saturation variables

- The solutions observed with the coherent theory (CT), finite difference (FD) and orthogonal collocation (OC) were not too dissimilar; they were however, not exact. The slight differences are due to assumed discrete values used for each method. The coherent wave fronts match the other two methods at their points of turning, where the effective wave dispersion is zero.

The simulation results of the predicted profiles by the coherent theory (CT), finite difference (FD) and orthogonal collocation (OC) techniques illustrate the typical effects of a pattern flood such as the early breakthrough of the oil bank and the long tail on the oil surfactant curves. The results show only significant deviations in some sections. The only significant difference beween the coherent theory (CT) and finite difference (FD) results is that the finite difference (FD) profiles are continous while the coherent profiles exhibit discontinous. More oscillations are evident in the orthogonal collocation (OC) solution profiles. This indicates that orthogonal collocation (OC) solution is sensitive to oscillation than other methods. This is particularly noticeable in the curves following breakthrough. The finite difference (FD) is considerably more dissipative and therefore the small oscillations are absent. Also, the coherent suppresses oscillation and is less dissipative. These findings are beyond the findings of Hankins and Harwell [20] . For, the work of Hankins and Harwell [20] did not show any of these new findings discussed here.

- Several simulations were made to evaluate the dependence of oil recovery on slug size for a given amount of injected surfactant. This is an important design factor and has been studied by many people for some years. It is generally thought that a large dilute slug will be better in a heterogeneous reservoir. This seems intuitively correct since a small slug would seem to be proportioned into the lowest permeabilities in such a small amount that little if any oil recovery from those parts of the reservoir could be expected. This is by itself would not be an unreasonable strategy provided portion of the reservoir which is the target only, The dependence of oil recovery on slug size for a given amount of injected surfactant or for a fixed product of slug size time's surfactant concentration indicates that the more surfactant injected the more oil recovered. So the effect of surfactant slug size is predictable. The effect of surfactant concentration is also predictable in the same manner as more oil is recovered for increase in surfactant concentration injected.

- The oil recoveries are attributed to the reduction of oil-water interfacial tension. Residual oleic phase saturation for the water-oil surfactant system depends on interfacial tension Oil initially at residual saturation is displaced by injection of surfactant solution. Through a complicated and highly non-linear function of the phase concentrations, the surfactant solution lowers the oil-water interfacial tension, and thus mobilizes the trapped oil. Three separate phases can form when such fluids displace oil-water mixtures. There are two regions of moderate change corresponding to two "fronts". These are the leading edge of the surfactant and the solubilization (or miscible) front as the concentrations jump to their injected values. Physically, this region corresponds to the very rapid increase in the relative permeability of the aqueous phase due to the decrease in interfacial tension. This is, of course, what the surfactant is designed to do, and is a physically desirable feature of the process. Oil trapped after water flooding is displaced by the injection of a slug of surfactant solution. The surfactant solution lowers the oil-water interfacial tension and, thus mobilizes the trapped oil. Therefore, residual oleic phase saturation for the water-oil-surfactant system depends on interfacial tension.

The complexities could not have been detected by using only the coherent technique as used by Hankins and Harwell [25] . This is a major accomplishment of this work. Not only was the discontinuities discovered by this work, it also provides an insight into the complex behaviour of enhanced oil recovery process.

We will now focus on the ultimate objective of this work, which is the volume of oil recovery achievable by our simulator.

Almost no change occurred was observed in the oil recovered as the slug size changed from one percentage of the pore volume to another, provided the salinity gradient was exactly the same for all runs. The implication is that slug size per se makes almost no difference over the range of typical slug sizes used in practice. Other factors such as the mobility control do matter and may be affected by the type of slug used and, from the point of view of chemical of injected surfactant.

The dependence of oil recovery on slug size for a given amount of injected surfactant indicates that the more surfactant injected the more oil is recovered. This is because an injected surfactant disperses into oil and water, then lowers the interfacial tension thereby mobilizing more immobile oil. This is continued until the surfactant is diluted or lost due to adsorption by the rock. For significant incremental oil recoveries, several orders of magnitude reduction is needed. Hence, large quantities of surfactant are required for reduction of the interfacial tension to produce the desired effect. So, the effect of surfactant slug size is forecastable. The effect of surfactant concentration is also predictable in the same way as more oil is recovered for increases in surfactant concentration injected.

The oil recovered in the reservoir Case A (hypothetical data) is lower than that of reservoir Case B (Nigerian oil data). The comparison shows that the breakthrough occurs much earlier for Case B, the ultimate recovery is obtained for a much smaller throughput of injected pore volume and sooner than for Case A.

CONCLUSIONS

The applicability of the simulator for the solution of the model equations of multiphase, multicomponent flow and transport in a reservoir has been demonstrated using orthogonal collocation solution. The results of the orthogonal collocation solution were compared with those of finite difference and coherence solutions. The results obtained using this methodology revealed certain features unobserved by previous investigators [25] . The results indicate that the concentration of surfactants (C_1, C_2) for orthogonal collocation appears to show more features than the predictions of the coherence solutions and finite difference. The reason for the difference is the subject of continuing study. It is unlikely that the coherence approach alone could ultimately accommodate the comple- xities required for a full field reservoir simulator.

It is obvious that the coherent routes for the compositions of adsorbing surfactants correspond to the simpler case of aqueous phase chromatography, with modified eigenvalue. This observation also holds for "shock" waves. The existence of a partially coherent solution makes the prediction of reservoir response much more straightforward. However, the "partially" coherent solution only exists if surfactants do not partition into the oleic phase and fractional flow is not a function of surfactant composition; should either occur, a globally coherent solution may still be found, but the solution is more complex and difficult to handle. There is no equivalence of partial coherence in the orthogonal and finite difference methods. In these lie the possibility of the differences in the concentration profiles predicted by the three numerical techniques. Again, the use of the orthogonal collocation and finite difference solution provides easier solution to future possible problems that may arise as the simulator is being used.

The dependence of oil recovery on slug size for a given amount of injected surfactant indicates that the more surfactant that is injected, the more oil is recovered. For significant incremental oil recoveries, several orders of magnitude interfacial tension reduction are needed. Hence, large quantities of surfactant are required to produce the desired effect of high oil recovery.

The predicted oil recovery for the Nigerian oil reservoir is higher than that predicted for the hypothetical reservoir data. The displacement is stable in the Nigerian oil reservoir due to the moderate value of the oil/water viscosity instead of the other reservoir with high value of oil/water ratio.

REFERENCES

1. Glimm, J., Lindquist, B., McBryan, O.A. , Plohr, B. and Yaniv, S. (1983) Front Tracking for Petroleum Reservoir Simulation. SPE Reservoir Simulation Symposium, 15-18 November 1983, San Francisco, Paper ID: SPE-12238-MS.

2. Ewing, R.E. , Russel , T.F. and Wheeler, M.F. (1983) Simulation of Miscible Displacement Using Mixed Methods and a Modified Method of Characteristics. SPE Reservoir Simulation Symposium, 15-18 November 1983, San Francisco, Paper ID: SPE-12241-MS. http://dx.doi.org/10.2118/12241-MS

3. Zheng, C. (1993) Extension of the Method of Characteristics for Simulation of Solute Transport in 3 Dimensions. Groundwater, 31, 456-465.http://dx.doi.org/10.1111/j.1745-6584.1993.tb01848.x

4. Helfferich, F.G. (1981) Theory of Multicomponent, Multiphase Displacement in Porous Media. Society of Petroleum Engineers Journal, 21, 51-62.http://dx.doi.org/10.2118/8372-PA

5. Lake, L.W. (1989) Enhanced Oil Recovery. Prentice-Hall, London.

6. Dezabala, E.F. , Vislocky, J.M. , Rubin, E. and Radke, C.J. (1982) A Chemical Theory for Linear Alkaline Flooding. Society of Petroleum Engineers Journal, 245-258.http://dx.doi.org/10.2118/8997-PA

7. Hirasaki, G.J. (1981) Application of the Theory of Multicomponent, Multiphase Displacement to Three-Component, Two-Phase Surfactant Flooding. Society of Petroleum Engineers Journal, 21, 191-204. http://dx.doi.org/10.2118/8373-PA

8. Pope, G.A. , Carey , G.F . and Sepehrnoori, K. (1984) Isothermal, Multiphase, Multicomponent Fluid Flow in Permeable Media. Part II: Numerical Techniques and Solution. In Situ, 8, 41-97.

9. Lake, L.W. and Helfferich, F.G. (1978) Cation Exchange in Chemical Flooding: Part 2. The Effect of Dispersion, Cation Exchange, and Polymer/Surfactant Adsorption on Chemical Flood Environment. Society of Petroleum Engineers Journal, 18, 435-444.http://dx.doi.org/10.2118/6769-PA

10. Helfferich, F.G. (1989) The Theory of Precipitation/Dissolution Waves. AIChE Journal, 35, 75-87.

11. Araque-Martinez, A. and Lake, L.W. (2000) Some Frequently Overlooked Aspects of Reactive Flow through Permeable Media. Industrial Engineering Chemistry Research, 39, 2717-2724. http://dx.doi.org/10.1021/ie990881m

12. Harwel, J.H. , Helfferich, F.G. and Schechter , R.S. (1982) Effect of Micelle Formation on Chromatographic Movement of Surfactant Mixtures. AIChE Journal, 28, 448-459.http://dx.doi.org/10.1002/aic.690280313

13. Helset, H.M. and Lake, L.W. (1998) Three-Phase Secondary Migration of Hydrocarbon. Mathematical Geology, 30, 637-660. http://dx.doi.org/10.1023/A:1022339201394

14. Cirpka, O.A. and Kitanidis, P.K. (2001) Transport of Volatile Compounds in Porous Media in the Presence of a Trapped Gas Phase. Journal of Contaminant Hydrology, 49, 263-285.http://dx.doi.org/10.1016/S0169-7722(00)00196-0

15. Patton , J.R. , Coats, K.H. and Colegrove, G.T. (1971) Prediction of Polymer Flood Performance. Society of Petroleum Engineers Journal, 72-84.http://dx.doi.org/10.2118/2546-PA

16. Fayers, F.J. and Perrine, R.L. (1959) Mathematical Description of Detergent Flooding in Oil Reservoirs. Transactions of the AIME, 216, 277-283.

17. Claridge, E.L. and Bondor, P.L . (1974) A graphical Method for Calculating Linear Displacement with Mass Transfer and Continuously Changing Mobilities. Society of Petroleum Engineers Journal, 609-618. http://dx.doi.org/10.2118/4673-PA

18. Larson, R.G. (1979) The Influence of Phase Behaviour on Surfactant Flooding. Society of Petroleum Engineers Journal, 411-422. http://dx.doi.org/10.2118/6774-PA

19. Lake, L.W. , Pope, G.A. , Carey , G.F . and Sepehrnoori, K. (1984) Isothermal, Multiphase, Multicomponent Fluid Flow in Permeable Media. Part I: Description and Mathematical Formulation. In Situ, 8, 1-40

20. Hankins, N.P. and Harwell, J.H. (1997) Case Studies for the Feasibility of Sweep Improvement in Surfactant-Assisted Waterflooding. Isothermal, Multiphase, Multicomponent Fluid Flow in Permeable Media. Part I: Description and Mathematical Formulation, 17, 41-62. http://dx.doi.org/10.1016 /S0920-4105(96)00055-1

21. Siggel, L., Santa, M., Hansch, M., Nowak, M., Ranft, M., Weiss, H., Hajnal, D., Schreiner, E., Oetter, G. and Tinsley, J. (2012) A New Class of Viscoelastic Surfactants for Enhanced Oil Recovery. SPE Improved Oil Recovery Symposium, 14-18 April 2012, Tulsa, Paper ID: SPE-153969-MS. http://dx.doi.org/10.2118/153969-MS

22. Xu, Y .H. and Lu, M. (2011) Microbially Enhanced Oil Recovery at Simulated Reservoir Conditions by Use of Engineered Bacteria. Journal of Petroleum Science and Engineering, 78, 233-238.

23. Leach, A . and Mason , C.F. (2011) Co-Optimization of Enhanced Oil Recovery and Carbon Sequestration. Resource and Energy Economics, 33, 893-912.http://dx.doi.org/10.1016/ j.reseneeco.2010.11.002

24. Harwell, J.H. (2012) Enhanced Oil Recovery Made Simple. Journal of Petroleum Technology, 60, 42-43.

25. Hankins, N.P. and Harwell, J.H. (2004) Application of Coherence Theory to a Reservoir Enhanced Oil Recovery Simulator. Journal of Petroleum Science and Engineering, 42, 29-55. http://dx.doi. org/10.1016/j.petrol.2003.10.002

26. Oyedeko, K.F. (2012) Design and Development of a Simulator for a Reservoir Enhanced Oil Recovery Process. Ph.D. Thesis, Lagos State University, Lagos.

27. Trogus, F.J. , Schecchter, R.S. , Pope, G.A. and Wade , W.H. (1979) A New Interpretation of Adsorption Maxima and Minima. Journal of Colloid and Interface Science, 70, 293-305.

28. Helfferich, F.G. (1986) Multicomponent Wave Propagation: Attainment of Coherence from Arbitrary Starting Conditions. Chemical Engineering Communications, 44, 275-285.http://dx.doi.org/10.1080/00986448608911360

29. Villadsen, J.V. and Stewart , W.E. (1967) Solution of Boundary Value Problems by Orthogonal Collocation. Chemical Engineering Science, 22, 1483-1501.http://dx.doi.org/10.1016/0009-2509(67)80074-5

30. Villadsen, J.V. and Stewart , W.E. (1968) Solution of Boundary value problems by orthogonal Collocation, Chemical Engineering Science, 23, 1515.http://dx.doi.org/10.1016/0009-2509(68)89064-5

Low Pressure Chemical Vapor Deposition of Nb and F Co-Doped TiO$_2$ Layer

Satoshi Yamauchi, Shouta Saiki, Kazuhiro Ishibashi, Akie Nakagawa, and Sakura Hatakeyaa

Department of Biomolecular Functional Engineering, Ibaraki University, Hitachi, Japan

ABSTRACT

Nb and F co-doped anatase TiO$_2$ layers were deposited by low pressure chemical vapor deposition (LPCVD) at pressure of 3 mtorr using titanium-tetra-iso-propoxide (TTIP), O$_2$ and NbF$_5$ as precursor, oxidant and dopant respectively. Resistivity beyond 100 Ωcm for undoped layer was decreased with increasing supply of the dopant and dependent on the supply ratio of O$_2$ to TTIP and decreased to 0.2 Ωcm by the optimization. X-ray fluorescent spectroscopy showed Nb-content in the layer was decreased with the O$_2$-supply ratio. X-ray photo-spectroscopy indicated that F substituted O-site in TiO_2 by O_2-supply but carbon-contamination and F missing substitution in the O-site were significantly increased by excess O_2-supply. Further, it was suggested that the substituted F played an important role to reduce resistivity without significant contribution of O-vacancies. XRD spectra showed F missing substitution in the O-site degraded the crystallinity.

INTRODUCTION

Nowadays, Indium-Tin-Oxide (ITO) is widely used for transparent conducting oxide (TCO) to fabricate flat panel displays, solar-cells and so on. However, the amount of Indium on earth is significantly low as shown by Clarke number. Therefore, the alternative TCO without Indium has been attractively studied, for example, Ga-doped ZnO (GZO), F-doped SnO_2 (FTO) etc. [1] [2] . TiO_2, which has been extensively investigated in view of photo-induced applications using the photo-catalytic reactions and the hydrophilicity [3] [4] in addition to dielectric applications using the high dielectric constant and optoelectronic applications by the high refractive index [5] [6] , is also candidate for TCO because of the wide bandgap about 3.2 eV. It is however easily recognized that the conductivity control of TiO_2 is more difficult than ZnO and Sn_2O since d-bands are included in the crystal. Therefore, precise control of the deposition condition using sufficient donor and the post-annealing required to control the conduction because not only the donor but also defect d-band due to Ti^{3+} generated by oxygen-deficiency contributes to the conduction. For the purpose, laser-ablation and reactive sputtering with the post-annealing have been recently studied by using Nb as a sufficient donor in anatase-TiO_2 [7] [8] . It is mentioned that oxygen-deficiency is required to enhance the electronic activation of the donor [9] , which indicates the Ti^{3+} reduced from Ti^{4+} in TiO_2 contributes to the conduction, however, it is considered that the oxygen-vacancies cause the large lattice distortion and the insufficient stability. In contrast, it was reported that F substituted O-site in TiO_2 reduces Ti^{4+} to Ti^{3+} [10] , in which the lattice distortion was significantly reduced comparing to oxygen-vacancy since the ion radius of F^- (0.136 nm) is close to O^{2-} (0.140 nm). It is therefore expected that F can play an important role in Nb-doped TiO_2 layer to increase the conductivity preventing the lattice distortion, but the study for Nb-F co-doping in TiO_2 has not been studied whereas F doping has been reported for the photo-activities [11] [12] .

Conductivity controlled TiO_2 layers are also expected to use for such as chemical sensors, solar-cells, and the other electronic devices in solutions because of the high resistance against acidand alkaline-electrolytes. For such device applications, chemical vapor deposition (CVD) is a useful process comparing to physical vapor deposition in the view of step-coverage on the three-dimensionally structured surfaces to enhance the sensitivity and the efficiency in addition to the micro-fabrication by reactive ion etching [13] . CVD of TiO_2 layer has been studied by using metalorganic precursor of titanium-tetra-iso-propoxide (TTIP: $Ti(O-i-C_3H_7)_4$ aiming at highefficient photo-induced properties, but few study for the conductivity control has been reported.

In this paper, low-pressure CVD (LPCVD) by TTIP and O_2 mixed gas was applied for TiO_2 deposition to control conductivity by Nb-F co-dope using NbF_5 as the dopant with the studies by X-ray fluorescent spectroscopy, X-ray photo-spectroscopy and X-ray diffraction.

EXPERIMENTAL

Deposition of TiO_2 Layer

A bell-jar type reactor with the base pressure under 1×10^{-5} torr by a combination of diffusion pump (D.P.) and a rotary pump (R.P.) as shown in Figure 1 was used for LPCVD of titanium-oxide. Titanium tetra-iso-propoxide (TTIP, $Ti(O-i-C_3H_7)_4$: 99.7%-purity) in liquid-phase was charged into a quartz cell and initially purified in vacuum at 50°C for 3 hrs, then vaporized at 70°C to introduce into the reactor through a stainless tube for TiO_2 deposition. The gas-phase TTIP was introduced without any carrier gas but pure oxygen gas (99.9999%-purity) was simultaneously introduced into the reactor through the other gas inlet. The supply ratio of O_2/TTIP was controlled by monitoring the reactor pressure using Shultz gage when the TTIP and the O_2 was individually introduced into the reactor. Niobium pentafluoride (NbF_5: 98%-purity) powder was charged in a crucible consists of boron-nitride (BN) and then thermally evaporated. 1 mm-thick quartz plate with optically flat surface was used as substrate, which was mounted on a substrate holder after chemical cleaning. Temperature of the substrate holder and the crucible were monitored by K-type thermo-couples (T.C.) and controlled by resistive heating and PID-systems.

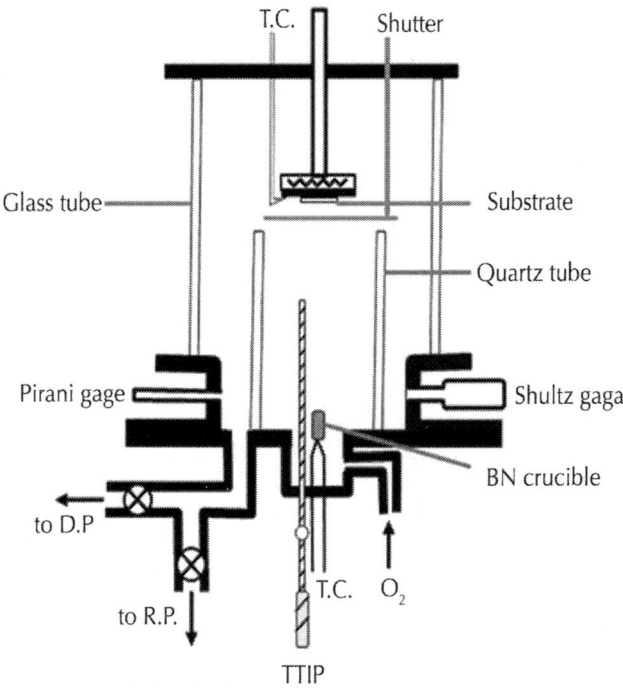

Figure 1: LPCVD apparatus for TiO$_2$ deposition.

Evaluation

Thickness of the layer was checked by a surface profiler (DEKTAK150). Resistivity was evaluated by Van Der Pauw (VDP) method. Densities and chemical states of impurities were analyzed by X-ray fluorescent spectroscopy (XRF: Shimazu XRF-1700) and X-ray photo-spectroscopy (XPS: Thermo VG Scientific,UK). Crystallographic behavior was examined by θ-2θ X-ray diffraction (RIGAKU: RAD-C) using CuKα.

RESULTS AND DISCUSSIONS

Dependence of Resistivity on the Deposition Condition

Figure 2 shows dependence of undoped-TiO_2 deposition rate on the deposition temperature in Arrhenius plot. The deposition rate was increased with the temperature by the activation energy of 138 kJ/mol then saturated above 360°C. The activation energy using TTIP and O_2 was seemed to be larger than that in other reports demonstrated at relatively high pressure above 0.1 torr [14] [15] but similar to the value of 150 kJ/mol by CVD using TTIP in N_2-gas with the high flow rate [16] . It has been considered that dissociation of TTIP is enhanced by oxygen, however, the result indicated TTIP was thermally dissociated in such low pressure according to the previously reported schemes as below [17] [18] ,

$$ Ti(OC_3H_7)_4 \rightarrow TiO_2 + 2C_3H_6 + 2HOC_3H_7 $$

Or

$$ Ti(OC_3H_7)_4 \rightarrow 2HOTi(OC_3H_7)_3 + 2C_3H_6 \rightarrow (OC_3H_7)_3 TiOTi(OC_3H_7)_3 + H_2O + 2C_3H_6 $$

In such feature, the deposition above 360°C can be recognized to be limited by TTIP-supply. In this work, all samples for the study of Nb-F doping were deposited at 380°C in the TTIP-supply limiting region for removal residual impurities such as carbon and/or hydrocarbons and to enhance crystallization into the anatase-phase.

Resistivity of the undoped-layer deposited at 380°C was higher than 100 Ωcm because the resistivity was over the evaluation-limit in our VDP-system which was able to evaluate the resistivity below 100 Ωcm. On the other, the resistivity of Nb-F doped TiO_2 layer was significantly reduced with the crucible temperature for NbF_5 evaporation (T_{NbF5}) and dependent on $O_2/(TTIP + O_2)$ gas supply ratio as shown in Figure 3. Figure 3(a) shows dependence of the resistivity (solid-circle and solid-line) and the deposition rate (open-circle and dot-line) on the T_{NbF5}, where the layers with the thickness about 200 nm were deposited by the $O_2/(TTIP + O_2)$ supply ratio of 0.50. The resistivity was decreased with the T_{NbF5} below 70°C and above 100°C then saturated about 1 Ωcm at the T_{NbF5} above 160°C, but increased in the range from 70 to 100°C. The increase of resistivity with the T_{NbF5} was due to melt of NbF_5 at the temperature above 70°C, that is, the NbF_5 was sublimated below 70°C and evaporated above 100°C after melting. It is noted that the carrier density and Hall mobility could not be determined by the VDP-system because of trap-induced conductions such as hopping, but conventional check by Zeebeck effect showed n-type conductivity for the doped layers. In addition to the decrease of resistivity, the deposition rate was decreased with NbF_5-supply as shown by open circles in Figure 3(a), which indicated the dopant influenced dissociation of TTIP. The feature for deposition rate influenced by NbF_5 was also dependent on O_2- supply ratio as shown by open-circles in Figure 3(b), where the NbF_5 was evaporated at 160°C. It is noted that total gas pressure of TTIP + O_2 for the various gas ratio was kept at 3 mtorr, therefore, the deposition rate of undoped layer was linearly decreased with the O_2-partial pressure according to decrease of TTIP-supply rate. However, the deposition rate of Nb-F doped layer was non-linearly decreased with the O_2-supply ratio, especially, above 0.35. For example, deposition rate of the doped layer deposited by the gas-supply ratio of 0.50 was 5 nm/min whereas the rate was 9 nm/min for undoped layer as shown in Figure 1. Since TTIP was thermally dissociated without oxidation, the obvious decrease of deposition rate for the doped layer could be recognized to be caused by reactive species dissociated from NbF_5 by the oxidation. On the other, the resistivity of the doped layer with the thickness about 200 nm was decreased with the O_2-supply ratio and as low as 0.2 Ωcm by the O_2-supply ratio of 0.38, but increased with the O_2-supply ratio above 0.40 and then saturated about 1 Ωcm. As a result, the O_2-supply for the deposition rate and the resistivity could be considered in three-regions such as Region I: poor O_2-supply (O_2 supply ratio under 0.38,) Region II; sufficient O_2-supply (O_2 supply ratio about 0.38), Region III; excess O_2-supply (O_2supply ratio beyond 0.38).

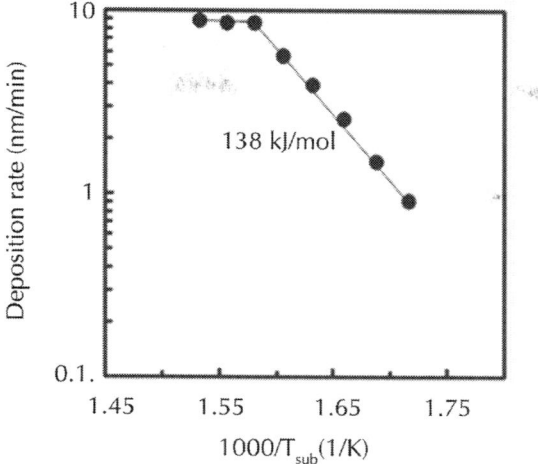

Figure 2: Arrhenius plot of TiO_2 deposition rate for the deposition temperature.

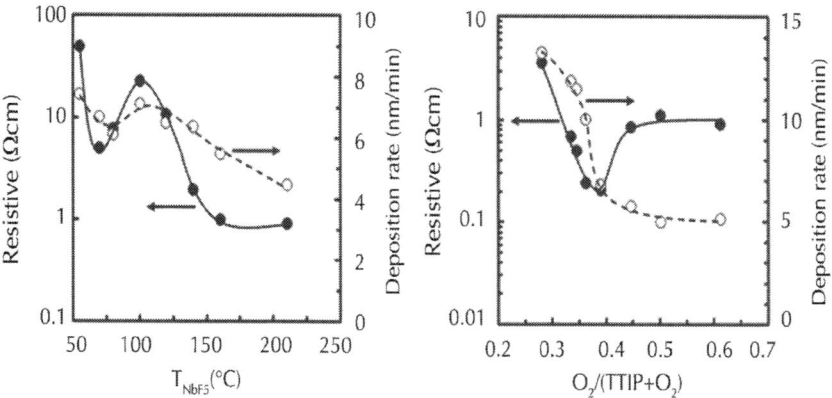

Figure 3: Resistivity (solid-circle and solid-line) and deposition rate (open-circle and dotline) of Nb-F doped TiO_2 layer (a) deposited by gas supply ratio of $O_2/(TTIP + O_2) = 1.0$ at various Nb-F crucible temperature ($TNbF_5$), and (b) deposited by NbF_5 supply at the $TNbF_5 = 160°C$ in various $O_2/(TTIP + O_2)$ supply ratio.

XRF and XPS Evaluations

Figure 4 shows XRF spectra of Nb-F doped layers deposited by the $O_2/$ (TTIP + O_2) supply ratio of 0.27 in Region I (green-line), 0.38 in Region II (black-line) and 0.61 in Region III (red-line) respectively where NbF_5 was evaporated at the T_{NbF5} of 160°C. Background in the spectra was numerically removed by poly-nominal function and then the spectra were normalized by the Ti-K_α intensity. The intensity due to Nb-K_α was decreased with the O_2-supply ratio, which resultantly indicated Nb-content was decreased with the gas supply ratio as shown in the inset. Because the deposition rate was decreased with the gas-supply ratio for same supply rate of NbF_5 (Figure 3(b)), the variation of Nb-content could not be simply discussed on the dopant-supply rate for the deposition rate. It is expected that $NbOF_x$ and F were formed from NbF_5 by the oxidation, and the density on the deposition surface were increased with O_2-supply rate. While Nb-content in the layer should be increased with the density of $NbOF_x$ on the surface because of the sticking probability higher than NbF_5, the experimental result in Figure 4 showed reversal dependence. Therefore, it is speculated that F dissociated from NbF_5 was an important species to clarify the doping, however, the content in the layers could not be evaluated by XRF because X-ray emission rate of the element was significantly low in addition to the low concentration in the layer. In contrast, the element could be studied on the chemical states by XPS. Figure 5 shows high resolution (a) C1s and (b) F1s XPS spectra of Nb-F doped TiO_2 layers deposited by the $O_2/$(TTIP + O_2) supply ratio of 0.27 (green-line), 0.38 (black-line) and 0.61 (red-line) respectively, with the C1s spectrum of undoped layer by the O_2-supply ratio of 0.50 (blue-line). It is noted that the spectrum originated from Nb was under detectable limit. Extremely weak C1s spectrum peak at 284.3 eV was appeared on the undoped layer, which indicated TTIP was successfully dissociated to TiO_2. On the other, significantly increased spectra were observed in the doped layers depending on the O_2-supply ratio, where the peak energy was shifted to 284.7 eV. Although the peak at 284.3 could be assigned to physisorbed carbons, the shift of peak-energy and the obvious increase of intensity indicated that the origin was not only adventitious carbons but also carbon-compounds included in the layers. Recently, Karlsson et al. studied TTIP-dissociation in ultra-high vacuum by XPS and showed C1s spectra peak at 285 eV is attributed to adsorbed methyl-group originated from TTIP [19] . Additionally, in the case of the layer deposited by the excess O_2-supply rate in Region III, spectrum in higher energy side was appeared at 288.5 eV. A lot of study for fluorinated hydro-carbons have been achieved by XPS and indicated that the spectra above 288 eV is originated from fluorinated carbon [20] [21] . It is consequently considered that alcohol group in TTIP was fluorinated in the condition of Region III, which was resulted in decrease of the deposition rate as shown in Figure 3 and increase of the residual methyl-group indicated by obvious increase of the spectrum peak at 284.7 eV. The fluorination was also observed in F1s spectra as shown in Figure 5(b). In the spectra, the most intense spectra peak at 684.5 could be attributed to adsorbed F ion or $TiOF_2$-F [22] . The spectrum peak at 688 assigned to C-F [23] was intense for the layer deposited by the excess O_2-supply rate (red-line), in which the intensity of C1s spectrum due to fluorinated carbon peak at 288.5 eV was also appeared. The other spectra peak at 689 eV and 690 eV could be assigned to F substituted in O-site of TiO_2 [24] [25] . The results indicated F could be doped in the O-site by O_2-supply in Region II and III. It is interesting that the result suggested the electronic activation efficiency of Nb in the layer was increased by the substituted F because resistivity of the layer deposited by the gas-supply ratio of 0.27 in Region I was significantly high comparing to the layer by the gas-supply ratio of 0.38 in Region II whereas the Nb-content was decreased with the gas-supply ratio.

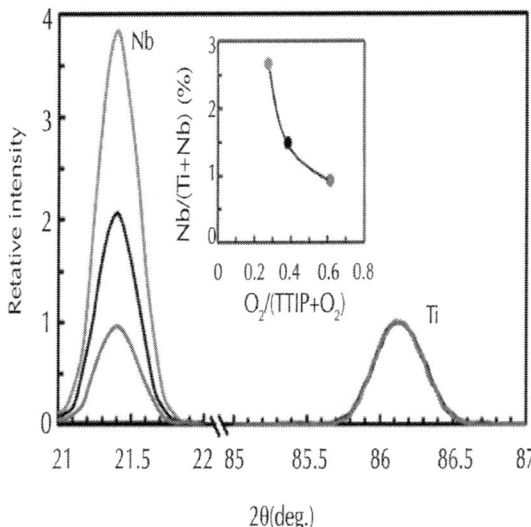

Figure 4: XRF spectra for Nb-K$_a$ (Nb) and Ti-K$_a$ (Ti) of Nb-F doped TiO$_2$ layers deposited by O$_2$/(TTIP + O$_2$) supply ratio of 0.27 (green-line), 0.38 (black-line) and 0.61 (red-line), where the intensity was normalized by the intensity of Ti-K$_a$. The inset shows the evaluated Nb-content.

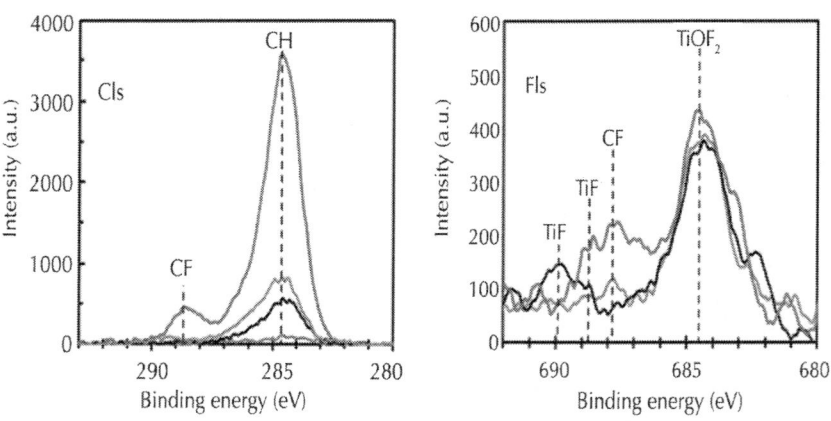

Figure 5: (a) C1s and (b) F1s XPS spectra of Nb-F doped TiO$_2$ layers deposited by various O$_2$/(TTIP + O$_2$) supply ratio of 0.28 (green-line), 0.38 (black-line) and 0.61 (red-line). The dot line in the (a) shows the spectrum of undoped TiO$_2$ layer deposited by the supply ratio of 0.50.

The role of oxygen has been in progress by first-principles molecular-orbital calculations, but it can be expected as follows.

NbF_5 is adsorbed on the deposition surface (probably to titanium) where H_2O formed during Ti-O-Ti bridging from Ti-OH---OH-Ti is simultaneously supplied to the NbF_5, then the NbF_5 is oxidized and HF is desorbed from the surface as the below scheme.

$$NbF_5 + nH_2O \rightarrow NbO_nF_{5-2n} + 2nHF \uparrow$$

In this scheme, F-content in the layer is decreased with H_2O-density on the surface.

NbF_5 adsorbed on the deposition surface is dissociated by oxygen as follow scheme, and then the dissociated F with the high reactivity adsorbs to titanium on the surface with high sticking probability or fluorinates the TTIP.

$$NbF_5 + mO_2 \rightarrow NbO_{2m}F_{5-2m} + 2mF \uparrow$$

The adsorbed F decreases the sticking probability of the next coming NbF_5 and TTIP, which resultantly decreases Nb-content in the layer and the deposition rate. Further, the fluorination disturbs thermal dissociation of TTIP, which is resulted in decreased of the deposition rate and increase of residual carbons in the layer. Of course, the dissociation of NbF_5 by H_2O as shown in the Scheme 1 is also including during the deposition but the oxygen-contribution becomes dominant by the high O_2-supply ratio according to the increased O_2-partial pressure comparing to H_2O on the surface.

Figure 6 shows the content ratio of F/Ti (solid-circle) and F/Nb (open-circle) in the layers for various O_2/(TTIP + O_2) supply ratio, where the ratio of F/Nb was estimated by the F/Ti ratio obtained by XPS and the Nb/Ti ratio evaluated by XRF. The F/Ti ratio was significantly increased by the high O_2-supply ratio in Region III. On the other, F was excessively contained in the layer than Nb (F/Nb > 1.0) and the F/Nb ratio was uniquely increased with the O_2-supply ratio. These results indicated that F was easily introduced into the layer by supporting of oxygen whereas the Nb-content was decreased with the O_2-supply, which was consistent with the result expected by the Sequence 1 and the Sequence 2.

Oxygen chemical state in the layer was also influenced by the O_2/ (TTIP + O_2) supply ratio. Figure 7shows O1s XPS spectra of Nb-F doped layers deposited by the gas-supply ratio of 0.27 (green-line), 0.38 (black-line) and 0.61 (red-line) with the spectrum of undoped layer by the gas ratio of 0.50 (blue-line), where the deconvo- luted spectra using Gaussian-function were also shown by dot-lines. Single spectrum peak at 529.5 eV were observed for the undoped and the doped layers deposited by the gas ratio of 0.27 and 0.38, but the other spectrum peak at 531.9 eV was also included for the doped layer by high O_2- supply ratio of 0.61. Previously, the double spectra was discussed for ITO layers, in which the spectrum at low high energy is concluded to be originated from oxygen neighboring cations coordinated six oxygens but the spectrum is shifted toward higher energy-side by contribution of oxygen-vacancies [26] . It is simply considered that oxygen-deficiency is reduced by increase of O_2-supply ratio. However, the results shown in Figure 7 indicated the oxygen-vacancies were significantly increased by high O_2-supply rate in Region III, which could be concluded that excess fluorine dissociated from NbF_5 caused the oxygen-vacancies. Commonly, annealing to reduce resistivity of Nb-doped TiO_2 layers has been processed in vacuum or reduction environment since oxygen-vacancies play an important role to reduce resistivity [27] [28] . In contrast, resistivity of the Nb-F doped layer was decreased with the O_2-supply rate in Region I and Region II as shown inFigure 3(b), but the density of oxygen-vacancies were not increased. The result apparently indicated fluorine substituted to oxygen in TiO_2 was important to reduce the resistivity of the Nb-F doped layer without the contribution of oxygen-vacancies. Further, the resistivity was increased in Region III whereas the density of oxygen-vacancy was significantly increased.

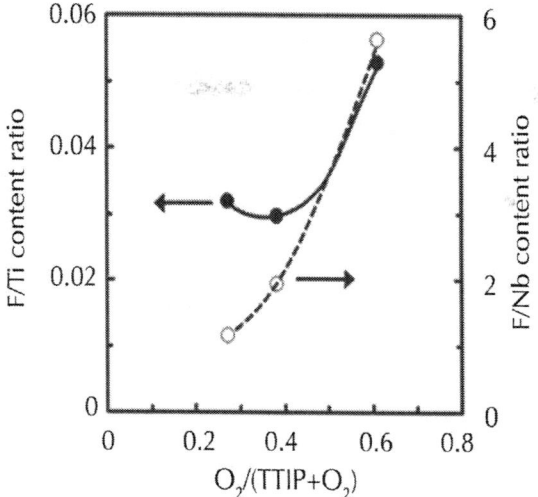

Figure 6: F/Ti content ratio evaluated by XPS (solid-circle and left-axis) and estimated F/Nb content ratio from XPS and XRF data (open-circle and right-axis) for $O_2/(TTIP + O_2)$ supply ratio.

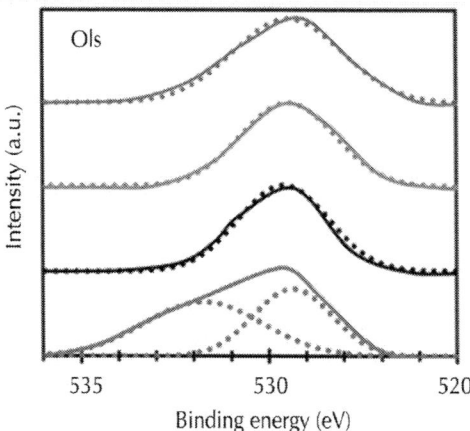

Figure 7: O1s XPS spectra of Nb-F doped TiO_2 layers deposited by various $O_2/(TTIP + O_2)$ supply ratio of 0.28 (green-line), 0.38 (black-line) and 0.61 (red-line) with the spectrum of undoped layer (blue-line) deposited by the gas supply ratio of 0.50. The dot lines show the deconvoluted spectra performed by Gaussian function.

Crystallographic Property

Figure 8 shows θ-2θ XRD spectra of Nb-F doped TiO_2 layers with the spectrum of undoped layer (blue-line) deposited by $O_2/(TTIP + O_2)$ supply ratio of 0.50, where the doped layers were deposited by the gas supply ratio of 0.27 (green-line), 0.38 (black-line) and 0.61 (red-line) with the dopant evaporation at 160°C. The spectra revealed the layers were poly-crystallized into the anatase-phase without the other phase such as rutile, brookite and the other compound as $TiOF_2$. The inset shows the spectra around anatase-TiO_2 (101) diffraction peak. The diffraction peak at $2\theta = 25.32°$ for the undoped layer was slightly shifted toward larger angle comparing to $\theta = 25.28°$ comparing to that of the bulk [29] , which indicated tensile stress was induced in the layer due to difference of thermal expansion coefficient between the layer and the quartz substrate. On the other, the peak angle was 25.29° for the layers deposited by the O_2-supply ratio of 0.27 and 0.38, and 25.25° for the layer deposited by the gas-supply ratio of 0.61. Previously, it was reported that the peak angle is decreased (the d-spacing is increased) with Nb-content in Nb-doped TiO_2 layer fabricated by sputtering deposition and the post-annealing [27] . In this work, it is believed the mismatch for the ion radius between Ti^{4+} (0.061 nm) and Nb^{5+} (0.064 nm) in the cation-site was compensated by fluorine ion radius (F^-: 0.136 nm) smaller than that of oxygen ion (O^{2-}: 0.140 nm) in anion-site. However, the peak angle for the doped layers was shifted toward low angle-side comparing to that of the undoped-layer. Since Nb-content in the layers was decreased with the O_2-supply ratio and F-content was significantly increased by the gas-supply ratio of 0.61 as shown in Figure 4 and Figure 6, the peak angle should be shifted toward high angle-side (decrease of the d-spacing) comparing to that of the layers deposited in Region I and Region II. It is therefore difficult to recognize that the shift was due to the substituted Nb and F and expected to be caused by the fluorine missing the substitution in the oxygen site without crystallization into $TiOF_2$. As a result, it could be mentioned that removal the insufficient fluorine is required to improve crystallinity for further reduction of the resistivity.

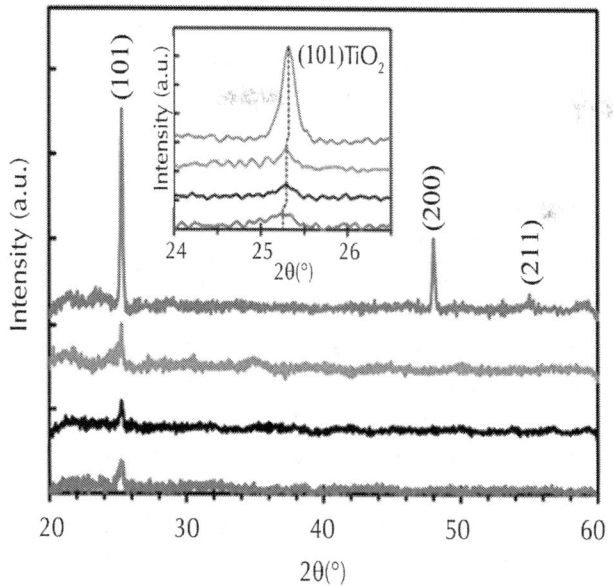

Figure 8: XRD spectra of undoped (blue-line) deposited by $O_2/(TTIP + O_2)$ supply ratio of 0.50 and Nb-F doped TiO_2 layers deposited by $O_2/(TTIP + O_2)$ supply ratio of 0.27 (green-line), 0.38 (black-line) and 0.61 (red-line), where NbF_5 was supplied at the crucible temperature of 160°C.

CONCLUSIONS

Nb and F were simultaneously doped in anatase-TiO_2 by using NbF_5 on low pressure chemical vapor deposition using TTIP and O_2. Nb and F-content in the layer were dependent on the O_2-supply ratio, in which the Nb-content was decreased with increasing the gas supply ratio but F was increased in the high gas-supply ratio. XPS studies indicated F substitutes O-site in TiO_2 by the O_2-supply but carbon contamination is also increased by the excess O_2-supply. It was suggested by comparison between the resistivity and the XPS results that F in the O-site plays an important role to reduce the resistivity without oxygen-vacancies. XRD spectra speculated F missing substitution in the O-site degrades the crystallinity.

REFERENCES

1. Assunção, V., Fortunato, E., Marques, A., Águas, H., Ferreira, I., Costa, M.E.V. and Martins, R. (2003) Influence of the Deposition Pressure on the Properties of Transparent and Conductive ZnO:Ga Thin-Film Produced by r.f. Sputtering at Room Temperature. Thin Solid Films, 427, 401-405. http://dx.doi.org/10.1016/S0040-6090(02)01184-7

2. Lewis, B.G. and Paine, D.C. (2000) Applications and Processing of Transparent Conducting Oxides. MRS Bulletin, 25, 22-27. http://dx.doi.org/10.1557/mrs2000.147

3. Wang, R., Hashimoto, K. and Fujishima, A. (1997) Light-Induced Amphiphilic Surfaces. Nature, 388, 431-432. http://dx.doi.org/10.1038/41233

4. Mills, A., Lepre, A., Elliott, N., Bhopal, A., Parkin, I.P. and Neill, S.A. (2003) Characterisation of the Photocatalyst Pilkington Activ™: A Reference Film Photocatalyst? Journal of Photochemistry and Photobiology A: Chemistry, 160, 213-224. http://dx.doi.org/10.1016/S1010-6030(03)00205-3

5. Campbell, S.A., Kim, H.S., Gilmer, D.C., He, B., Ma, T. and Gladfelter, W.L. (1999) Titanium Dioxide (TiO_2)-Based Gate Insulators. IBM Journal of Research and Development, 43, 383-392. http://dx.doi.org/10.1147/rd.433.0383

6. Martinet, C., Paillard, V., Gagnaire, A. and Joseph, J. (1997) Deposition of SiO_2 and TiO_2 Thin Films by Plasma Enhanced Chemical Vapor Deposition for Antireflection Coating. Journal of Non-Crystalline Solids, 216, 77-82. http://dx.doi.org/10.1016/S0022-3093(97)00175-0

7. Hitosugi, T., Ueda, A., Furubayashi, Y., Hirose, Y., Konuma, S., Shimada, T. and Hasegawa, T. (2006) Fabrication of TiO_2-Based Transparent Conducting Oxide Films on Glass by Pulsed Laser Deposition. Japanese Journal of Applied Physics, 46, L86-L88. http://dx.doi.org/10.1143/JJAP.46.L86

8. Gillispie, M.A., van Hest, M.F.A.M., Dabney, M.S., Perkins, J.D. and Ginley, D.S. (2007) rf Magnetron Sputter Deposition of Transparent Conducting Nb-doped TiO_2 Films on $SrTiO_3$. Journal of Applied Physics, 101, 033125-1-4. http://dx.doi.org/10.1063/1.2434005

9. Hoang, N.L., Yamada, N., Hitosugi, T., Kasai, J., Nakao, S., Shimada, T. and Hasegawa, T. (2008) Low-temperature Fabrication of Transparent Conducting Anatase Nb-doped TiO_2 Films by Sputtering. Applied Physics Express, 1, 115001-115003. http://dx.doi.org/143/APEX.1.115001

10. Di Valentin, C., Pacchioni, G. and Selloni, A. (2009) Reduced and n-Type Doped TiO_2: Nature of Ti^{3+} Species. Journal of Physical Chemistry C, 113, 20543–20552. http://dx.doi.org/10.1021/jp9061797

11. Li, D., Haneda, H., Hishita, S., Ohashia, N. and Labhsetwar, N.K. (2005) Fluorine-doped TiO_2 Powders Prepared by Spray Pyrolysis and Their Improved Photocatalytic Activity for Decomposition of Gas-phase Acetaldehyde. Journal of Fluorine Chemistry, 126, 69-77. http://dx.doi.org/10.1016/j.jfluchem.2004.10.044

12. Di Valentin, C., Finazzi, E., Pacchioni, G., Selloni, A., Livraghi, S., Czoska, A.M., Paganini, M.C. and Giamello, E. (2008) Density Functional Theory and Electron Paramagnetic Resonance Study on the Effect of N-F Codoping of TiO_2. Chemistry of Materials, 20, 3706-3714. http://dx.doi.org/10.1021/cm703636s

13. Norasetthekul, S., Park, P.Y., Baik, K.H., Lee, K.P., Shin, J.H., Jeong, B.S., Shishodia, V., Lambers, E.S., Norton, D.P. and Pearton, S.J. (2001) Dry Etch Chemistries for TiO_2 Thin Films. Applied Surface Science, 185, 27-33. http://dx.doi.org/10.1016/S0169-4332(01)00562-1

14. Yang, W. and Wolden, C.A. (2006) Plasma-Enhanced Chemical Vapor Deposition of TiO_2 Thin Films for Dielectric Applications. Thin Solid Films, 515, 1708-1713. http://dx.doi.org/10.1016/j.tsf.2006.06.010

15. Ahn, K.H., Park, Y.B. and Park, D.W. (2003) Kinetic and Mechanistic Study on the Chemical Vapor Deposition of Titanium Dioxide Thin Films by In Situ FT-IR using TTIP. Surface Coating and Technology, 171, 198-204. http://dx.doi.org/10.1016/S0257-8972(03)00271-8

16. Yokozawa, M., Iwasa, H. and Teramoto, I. (1968) Vapor Deposition of TiO_2. Japanese Journal of Applied Physics, 7, 96-97. http://dx.doi.org/10.1143/JJAP.7.96

17. Fictorie, C.P., Evans, J.F. and Gladfelter, W.L. (1994) Kinetic and Mechanistic Study of the Chemical Vapor Deposition of Titanium

Dioxide Thin Films using Tetrakis-(Isopropoxo)-Titanium (IV). Journal of Vacuum Science & Technology A, 12, 1108-1113. http://dx.doi.org/10.1116/1.579173

18. Chen, S., Mason, M.G., Gysling, H.J., Paz-Pujalt, G.R., Blanton, T.N., Castro, T., Chen, K.M., Fictorie, C.P., Gladfelter, W.L., Franciosi, A., Cohen, P.I. and Evans, J.F. (1993) Ultrahigh Vacuum Metalorganic Chemical Vapor Deposition Growth and In Situ Characterization of Epitaxial TiO_2 Films. Journal of Vacuum Science & Technology A, 11, 2419-2429. http://dx.doi.org/10.1116/1.578587

19. Karlsson, P.G., Richter, J.H., Andersson, M.P., Johansson, M.K.-J., Blomquist, J., Uvdal, P. and Sandell, A. (2011) TiO_2 Chemical Vapor Deposition on Si(111) in Ultrahigh Vacuum: Transition from Interfacial Phase to Crystalline Phase in the Reaction Limited Regime. Surface Science, 605, 1147-1156. http://dx.doi.org/10.1016/j.susc.2011.03.001

20. Schmidt, S., Goyenola, C., Gueorguiev, G.K., Jensen, J., Greczynski, G., Ivanov, I.G., Czigány, Zs. and Hultman, L. (2013) Reactive High Power Impulse Magnetron Sputtering of CF_x Thin Films in Mixed Ar/CF_4 and Ar/C_4F_8 Discharges. Thin Solid Films, 542, 21-30.http://dx.doi.org/10.1016/j.tsf.2013.05.165

21. Crassous, I., Groult, H., Lantelme, F., Devilliers, D., Tressaud, A., Labrugère, C., Dubois, M., Belhomme, C., Colisson, A. and Morel, B. (2009) Study of the Fluorination of Carbon Anode in Molten KF-2HF by XPS and NMR Investigations. Journal of Fluorine Chemistry, 130, 1080-1085. http://dx.doi.org/10.1016/j.jfluchem.2009.07.022

22. Liu, G., Sun, C., Cheng, L., Jin, Y., Lu, H., Wang, L., Smith, S.C., Lu, G.Q. and Cheng, H.-M. (2009) Efficient Promotion of Anatase TiO_2 Photocatalysis via Bifunctional Surface-Terminating Ti-O-B-N Structures. The Journal of Physical Chemistry C, 113, 12317-12324. http://dx.doi.org/10.1021/jp900511u

23. Chiang, C.Y., Reddy, M.J. and Chu, P.P. (2004) Nano-Tube TiO_2 Composite $PVdF/LiPF_6$Solid Membranes. Solid State Ionics, 175, 631-635.http://dx.doi.org/10.1016/j.ssi.2003.12.039

24. Yang, G., Wang, T., Yang, B., Yan, Z., Ding, S. and Xiao, T. (2013) Enhanced Visible-Light Activity of F-N CoDoped TiO_2

Nanocrystals via Nonmetal Impurity, Ti^{3+} Ions and Oxygen Vacancies. Applied Surface Science, 287, 135- 142.http://dx.doi.org/10.1016/j.apsusc.2013.09.094

25. Li, Y., Jiang, Y., Peng, S. and Jiang, F. (2010) Nitrogen-Doped TiO_2 Modified with NH_4F for Efficient Photocatalytic Degradation of Formaldehyde under Blue Light-Emitting Diodes. Journal of Hazardous Materials, 182, 90-96.http://dx.doi.org/10.1016/j.jhazmat.2010.06.002

26. Fan, J.C.C. and Goodenough, J.B. (1977) X-Ray Photoemission Spectroscopy Studies of Sn-Doped Indium-Oxide Films. Journal of Applied Physics, 4B, 3524-3531.http://dx.doi.org/10.1063/1.324149

27. Sato, Y., Akizuki, H., Kamiyama, T. and Shigesato, Y. (2008) Transparent Conductive Nb-Doped TiO_2 Films Deposited by Direct-Current Magnetron Sputtering using a TiO_{2-x} Target. Thin Solid Films, 516, 5758-5762. http://dx.doi.org/10.1016/j.tsf.2007.10.047

28. Yu, C.-F., Sun, S.-J. and Chen, J.-M. (2014) Magnetic and Electrical Properties of TiO_2:Nb Thin Films. Applied Surface Science, 292, 773-776.http://dx.doi.org/10.1016/j.apsusc.2013.12.047

29. Weisssmann, S., et al. (1978) Selected Powder Diffraction Data for Metals and Alloys. JCPDS, Card No. 21-1272, 263.

Optimization of the Degradation of Hydroquinone, Resorcinol and Catechol Using Response Surface Methodology

Noureddine Elboughdiri[1, 2], Ammar Mahjoubi[1, 2],
Ali Shawabkeh[1], Hussam Elddin Khasawneh[1], and
Bassem Jamoussi[3]

[1]Department of Chemical Engineering, College of Engineering, University of Hail, Hail, KSA

[2]Department of Chemical Engineering Process, National School of Engineering of Gabes, Gabes, Tunisia

[3]Department of Physical, High Institute of Education and Continuing Formation, Bardo, Tunisia

ABSTRACT

A clay catalyst (montmorillonite and kaolinite) was prepared and used to degrade three phenolic compounds: hydroquinone, resorcinol and catechol obtained from the treatment the Olive Mill Wastewater (OMW) generated in the production of olive oil. The operating conditions of the degradation of these compounds are optimized by the response surface methodology (RSM) which is an experimental design used in process optimization studies. The results obtained by the catalytic tests and analyses performed by different techniques showed that the modified montmorillonites have very interesting catalytic, structural and textural properties; they are more effective for the catalytic phenolic compound degradation, they present the highest specific surface and they may support iron ions. We also determined the optimal degradation conditions by tracing the response surfaces of each compound; for example, for the catechol, the optimal conditions of degradation at pH 4 are obtained after 120 min at a concentration of H_2O_2 equal to 0.3 M. Of the three phenolic compounds, the kinetic degradation study revealed that the hydroquinone is the most degraded compound in the least amount of time. Finally, the rate of the catalyst iron ions release in the reaction is lower when the Fe-modified montmorillonites are used.

INTRODUCTION

All over the world, domestic and industrial discharges and agricultural pollution have greatly contributed to deteriorating surface and groundwater quality. These pollutants have an important impact on the environment and human health. Significant efforts have been made to reduce pollutant discharges by promoting clean technologies in industrial sectors and by cleaning up waste gas and waste water before discharging them into the environment [1].

The latest advances in water treatment have been achieved in the oxidation of organic compounds [2] [3] including phenolic compounds which are harmful to the environment and to human health [4] [5].

For example, the Advanced Oxidation Process (AOP) is based on the in situ formation of highly reactive chemical entities such as hydroxyl radicals (OH ⋅). These radicals possess a high power of oxidation of organic molecules to CO_2 and H_2O [6] compared to conventional oxidants like Cl_2, ClO_2 or O_3. These radicals are able to mineralize, partially or totally, most phenolic compounds. The AOP includes oxidation processes in the following:

- Homogeneous phase: Fe^{2+} $(Fe^{3+})/H_2O_2$, O_3/OH^- and O_3/H_2O_2 [7] - [9] ,
- Photochemical processes: UV alone, H_2O_2/UV, $H_2O_2/Fe^{2+}(Fe^{3+})/UV$, photo-Fenton, TiO_2/UV, O_3/UV [10] - [14] ,
- Direct electrochemical processes [15] [16] ,
- Indirect electrochemical processes [17] - [20].

Concerning the Fenton reagent (Fe^{2+} $(Fe^{3+})/H_2O_2$), applications is limited and they are destined specifically for industrial waste water treatment and contaminated soils depollution. The main limitations of this system include the need to operate at acidic pH (pH approximately 3), the need to use large quantities of chemical reagents and the formation of ferric hydroxide [21].

By using iron supported on clays, through different procedures such as impregnation, intercalation or insertion, in the catalytic oxidation, the effectiveness of this system is significantly increased and the doses of reagents are reduced [1] [22] . This study has several objectives. Initially, this study aims to improve the degradation of three phenolic compounds, namely, hydroquinone, resorcinol and catechol using Fenton-type catalysts. This study also aims to optimize the degradation operating conditions of the response surface methodology (RSM) and to study the degradation kinetics by three types of catalyst (pillared montmorillonite, intercalated montmorillonite and impregnated kaolinite). Finally the study investigates the catalytic performance of solid catalysts by characterizing them with different techniques.

EXPERIMENTAL

Materials and Instrumentation

Two commercial clay products, montmorillonite K10 and kaolinite powder were provided by Aldrich and Parachimic, respectively. Iron (III) solution was prepared by dissolving iron hydrochloric hexahydrate

(FeCl$_3$, 6H$_2$O; 99%; Merck) in purified water and for the catalytic test hydrogen peroxide solution (>30 wt%, CHEMI- PHARMA) was used.

The phenolic compounds were analysed by a standard HPLC instrument Younglin Acme 9000 (Eclipse column type C-18 5 μm 4.6 × 250 mm, wavelength: 254 nm, at an injection volume of 20 μL, mobile phase: methanol-water: 60/40). XRD (X'PERT Pro Philps Analytical diffractometer), surface area analysis by volumetric adsorption of nitrogen (ASAP 2000 Micromeritics), and atomic absorption spectroscopy (Novo 400 Analytikjena) were performed to characterize the resultant powders.

Catalyst Preparation

We used two types of clay to prepare the catalysts: a swelling (montmorillonite) and a non-swelling clay (kaolinite) which were prepared according to protocols I and II (Figure 1 and Figure 2) corresponding to the intercalation of montmorillonite, its bridging and the impregnation of kaolinite.

Catalytic Conditions

The catalytic test is done by preparing phenolic compounds solutions ([phenolic compound] = 50 mol·L^{-1}) changing the pH values (the pH were adjusted with NaOH or HCl) and adding Fe-powder (0.3 g) to the different prepared solutions using magnetic stirring to maintain a uniform suspension.

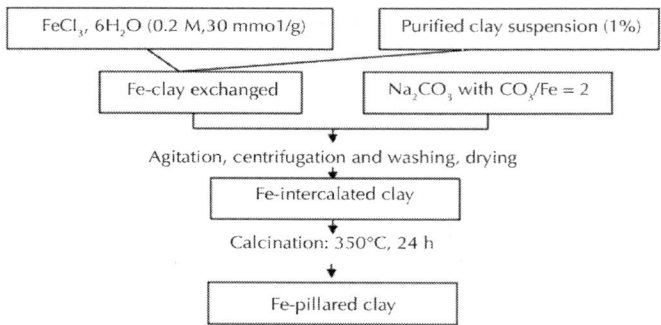

Figure 1: Preparation protocol of Fe-modified montmorillonite.

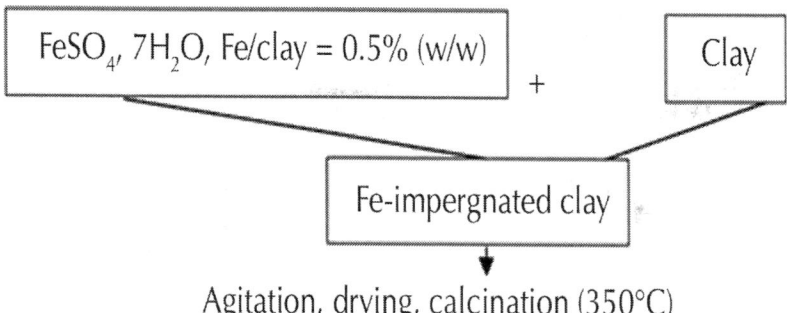

Figure 2: Preparation protocol of Fe-impregnated kaolinite.

After 10 min, varying concentrations of H_2O_2 were added into the reactor and time logged.

Investigation of Iron Release Conditions

After the three phenolic compounds degradation test ([phenolic compound] = 50 mol·L^{-1}, ambient temperature and pressure, pH = 6, catalyst mass = 0.3 g, 2 mL of [H_2O_2] = 0.26 M, solution volume = 200 mL), the solution was recovered and analyzed by atomic absorption to determine the amount of iron leached from the catalysts.

RESULT AND DISCUSSION

Optimization of Degradation of Phenolic Compounds by RSM

The general practice for optimizing the operating conditions for this process consists of varying one parameter and keeping the other ones constant. The major disadvantage of this single variable optimization is the disregard of interactive effects between the variables. In order to overcome this problem, optimization studies have been carried out using RSM. The theory behind RSM has been reviewed [23] [24].

This technique is an integration of experimental strategies, mathematical methods and statistical inference to determine the optimal level giving the most interesting response. RSM reduces the number of experimental trials needed to evaluate multiple parameters and their interactions and is therefore less laborious and time consuming than other approaches. RSM has been widely applied for optimizing processes in different domains such as the chemical process, geotechnical engineering and animal science research [25] - [27]. The phenolic compounds degradation is governed by several physicochemical factors. The selected factors and their corresponding ranges can be found in Hamzaoui, Jamoussi, M'nif and Missaoui Ines [28] [29].

Central Composite Design

The applied optimization approach is based on a central composite design and RSM, this method is one of the most important experimental designs used in process optimization studies [30]. In order to describe the nature of the response surface in the optimum region, a central composite design with five coded levels was performed (−1.68, −1, 0, 1, 1.68) and three factors (t: time (min), C: concentration of H_2O_2 (mol·L^{-1}) and pH)) were selected and processed simultaneously through the central composite design.

In general, central composite designs need a total of (2^k + 2k + N_0) runs where k is the number of studied factors, 2^k are the points from the factorial design, 2k the face-center points and N_0 the number of experiments carried out at the center (Figure 3). As usual, the experiments were carried out in random order to minimize the effect of systematic errors.

The parameter levels and coded values are given in Table 1. Fifteen experiments were performed corresponding to the three variables central composite design (Table 2). An example of the graphical representations of the distribution of these experimental points is given in Figure 4 [30]. The measured response was defined as degradation rate in %.

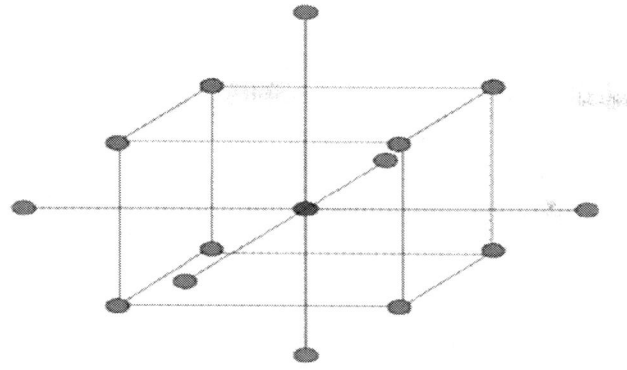

Figure 3: Spatial representation of a complete centred composite design with three factors.

Table 1: Parameter levels and coded values used in the experimental design

Parameters	Code	+α = +1.68	+1	0	−1	−α = −1.68
Time (min)	X_1	12.8	40	80	120	147
Concentration of H_2O_2 (mol·L^{-1})	X_2	0	0.02	0.11	0.2	0.26
pH	X_3	0.64	2	4	6	7.36

Table 2: Experimental design and response value

Experimental number	Time (min)	Concentration of H_2O_2(mol·L^{-1})	pH	Degradation rate (%) of		
				Hydroquinone	Resorcinol	Catechol
1	120	0.2	6	85.66	0.86	3.92
2	40	0.2	6	86.73	4.16	28.48
3	120	0.02	6	79.50	0	84.08
4	40	0.02	6	100.00	0	86.35
5	120	0.2	2	60.31	0	18.10
6	40	0.2	2	77.21	50.12	15.40
7	120	0.02	2	57.74	39.89	3.62

8	40	0.02	2	100.00	60.43	53.83
9	80	0.11	4	58.10	33.96	48.55
10	147	0.11	4	51.10	25.81	48.51
11	12.8	0.11	4	53.57	33.73	48.63
12	80	0.26	4	67.82	0	10.36
13	80	0	4	60.20	69.20	67.82
14	80	0.11	0.64	60.82	44.30	64.71
15	80	0.11	7.36	72.02	48.75	22.33

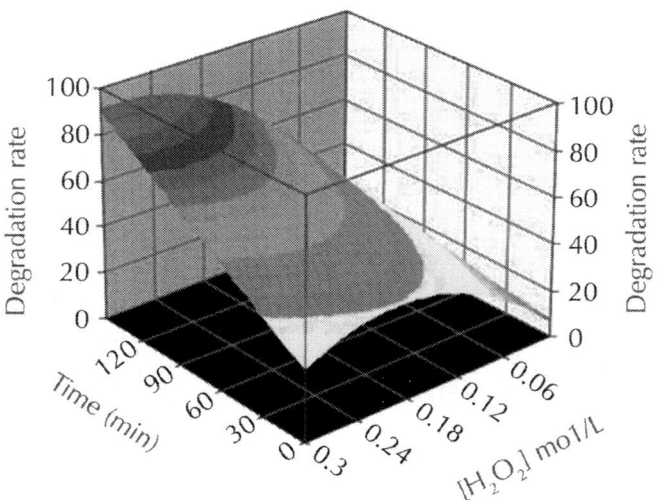

Figure 4: Three-dimensional plots for catechol degradation as a function of $[H_2O_2]$ and time, pH = 4.

Determination of Optimal Conditions

The quadratic response surface for the three factors involved generates a four-dimensional response surface, which can be illustrated in a three-dimensional (3D) response surface. For this, we used the software MAPLE 9.5. The surfaces were obtained by varying an experimental factor each time while keeping the third constant at its central value.

The visualization of these surfaces allows studying of the influence of each parameter and determining the optimal operating conditions to degrade the phenolic compounds, the subject of this study. An example of a graphical representation of the experimental points distribution of catechol degradation as a function of $[H_2O_2]$ and time at Ph = 4 is given in Figure 4. The measured response was defined as the degradation of phenolic compound in %.

Characterization of Fe-Modified Clays

Textural Characterization

Measurement of specific surface S_{BET} (Brunauer, Emmett and Teller method): We noted that the natural or modified montmorillonite presents a surface with high specificity compared to that of the kaolinite (Table 3).

This can be explained by the fact that the montmorillonite is swelling clay and its surface is formed by the internal basal surface, the external basal surface, and the external lateral surface, in addition to the surface of the pillars. This is unlike the non-swelling kaolinite which is characterized only by an external basal surface and an external lateral surface as shown in Figure 5 [31].

We also found that the S_{BET} of intercalated montmorillonite doubled from 142 to 327 m^2/g compared to that of natural montmorillonite which is probably due to the intercalation of iron polycations between the clay layers. As for the S_{BET} of pillared montmorillonite, we observed that it is less than the intercalated one. These results corroborate those obtained by Baccar, Batis and Ghorbel [32] who attributed this decrease to the formation of pillars of iron oxides after calcination.

Structural Characterization

Chemical analysis: Table 4 shows that the impregnated kaolinite contains a small amount of iron because the iron is impregnated on a low specific surface, in contrast to montmorillonite in which the iron can be set between its layers. In addition, the high quantity of iron in the two modified montmorillonites revealed that both clays were intercalated by iron ions. The mass percentage of iron (42.41% w/w) in pillared montmorillonite is higher than in the intercalated montmorillonite (33.7% w/w). This is probably due to the loss of water by intercalated montmorillonite therefore the mass percentage of iron increases relative to the total mass of the calcined clay.

X-ray diffraction: Figure 6 shows the XRD patterns of the Fe-modified montmorillonite.

The X-ray diffractogram of intercalated montmorillonite presents the 001 ray (peaks due to regular stacking of montmorillonite layers) whose position corresponds to a distance between the layers of $d001 = 10.24$ Å. This ray on the diffractogram of pillared montmorillonite moves to $d001 = 9.12$ Å.

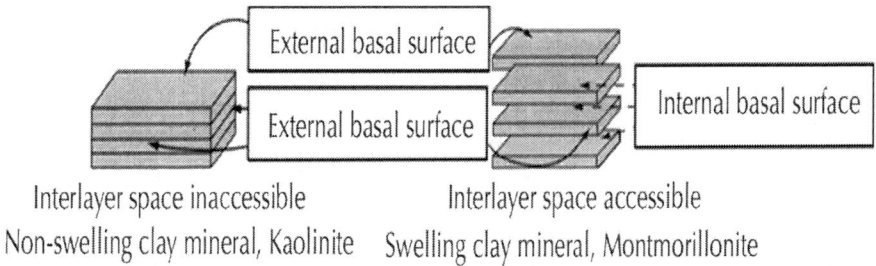

Interlayer space inaccessible   Interlayer space accessible
Non-swelling clay mineral, Kaolinite   Swelling clay mineral, Montmorillonite

Figure 5: Structure of kaolinite and montmorillonite.

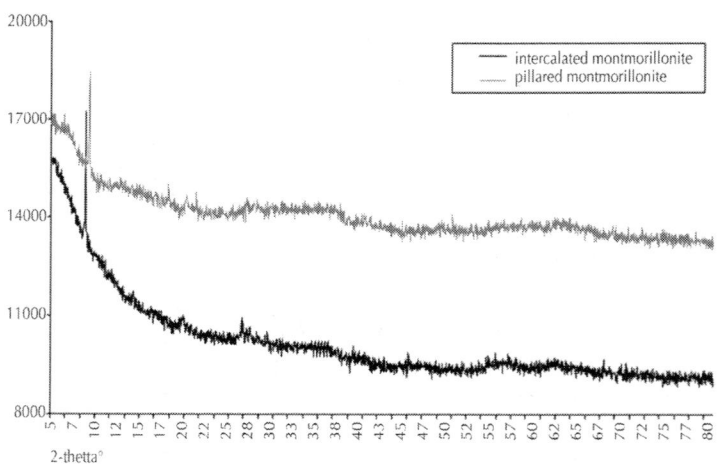

Figure 6: X-ray diffraction of Fe-modified montmorillonite.

Table 3: Specific surface area values

Samples	S_{BET} (m²/g)
Natural montmorillonite	142
Intercalated montmorillonite	327
Pillared montmorillonite	290
Natural kaolinite	11
Impregnated kaolinite	11

Table 4: Values of iron amount in clay

Samples	Intercalated montmorillonite	Pillared montmorillonite	Impregnated kaolinite
Iron amount (% w/w)	33.7	42.41	7.99

This slight decrease from 10.24 to 9.12 Å in the distance between the layers in the two clays reveals that the large polycations of iron inserted between the layers after calcination become pillars of iron oxides which are less bulky.

The X-ray diffractogram of the Fe-impregnated kaolinite (Figure 7) shows the rays of iron oxide that confirms their setting on kaolinite.

Investigation of Iron Release

We notice in Table 5 that the presence of leached iron in the solution induces the degradation of phenolic compounds according to homogeneous catalysis in addition to the heterogeneous catalysis. In fact the first catalysis is the result of the leached iron in the solution and the second one is caused by iron supported by clays.

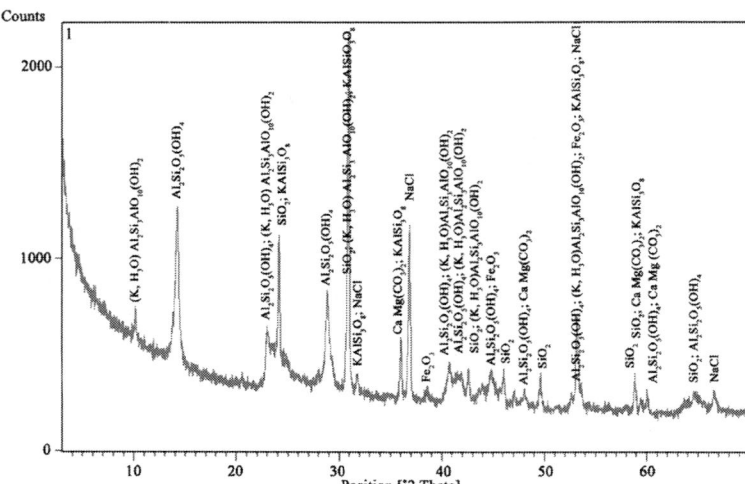

Figure 7: X-ray diffraction of Fe-impregnated kaolinite.

Table 5: Iron amount leached

Samples	Intercalated montmorillonite	Pillared montmorillonite	Impregnated kaolinite
Iron amount leached (mol·L^{-1})	0.415	1.619	58

We also note that the kaolinite leached a significant amount of iron (58 mg·L^{-1}). Thus, it does not yet support iron ions on its compact surface and these are not trapped between its layers. Both prepared montmorillonites released lesser amounts of iron (1.69 and 0.415 mg·L^{-1}) because iron ions were trapped between layers of clay. In summary, the catalysts prepared by swelling clays such as montmorillonite are more efficient than those prepared with compact structure clays.

Kinetics of Catalytic Degradation of Phenolic Compounds

The apparent degradation rate was determined by plotting the curve of ln ([C]$_0$/[C]), as a function of time. [C]$_0$ is the phenolic compound concentration at t = 0 and the [C] is the phenolic compound concentration at t. The equation was found to be first order.

The phenolic compounds degradation rate constants histogram (Figure 8) reveals that hydroquinone is the most degraded compound in the least amount of time and that modified montmorillonite is more efficient to degrade the phenolic compound than impregnated kaolinite; this is probably due to its interlayer structure which maintains iron ions between layers and to its high surface specificity compared to kaolinite.

Determination of Optimal Degradation Conditions

We traced the response surfaces of each compound by repeatedly varying two experimental factors and keeping the third constant at its central value (the pH was chosen at 4, the central value of time was 80 min and the H_2O_2 concentration was 0.11 M). For catechol (Figure 4), the optimal degradation conditions at pH 4 are obtained after 120 min with a concentration of H_2O_2 equal to 0.3 M. If we fix the H_2O_2 concentration at 0.11 M, the optimal conditions are obtained after 120 min and at pH less than 1.6. When the time of reaction is equal to 80 min, the optimal degradation is obtained at pH less than 3.2 and at a H_2O_2 concentration more than 0.12 M.

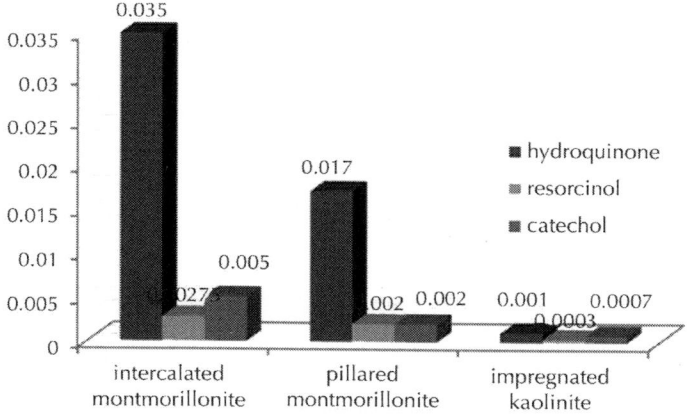

Figure 8: Histogram of rate constants of phenolic compounds degradation.

CONCLUSIONS

This study clearly shows the following:

- The use of the RSM allows the visualization of degradation response surfaces of each compound, the study of the influence of each parameter (pH, H_2O_2 concentration and time reaction) and the determination of the optimal operating conditions to degrade the selected phenolic compounds.

- The results obtained from the catalytic tests and analysis performed by different techniques showed that the modified montmorillonites have very distinct catalytic, structural and textural properties; they are more effective for the catalytic phenolic compound degradation, they present the highest surface specificity and they may support iron ions.

- The kinetic degradation of three phenolic compounds supports the fact that hydroquinone is the most degraded compound in the least amount of time.

- The control of the release of iron ions by the catalysts performed by atomic absorption spectroscopy shows that the rate of the catalyst iron ions release in the reaction when the Fe-modified montmorillonites are used is lower than when Fe-modified kaolinite is used.

REFERENCES

1.   Liotta, L.F., Gruttadauria, M., Carlo, G.D., Perrini, G. and Librando, V. (2009) Heterogeneous Catalytic Degradation of Phenolic Substrates: Catalysts Activity. Journal of Hazardous Materiels, 162, 588-606. http://dx.doi.org/10.1016/j.jhazmat.2008.05.115

2.   Hammami, S., Oturan, N., Bellakhal, N., Dachraoui, M. and Oturan, M.A. (2007) Oxidative Degradation of Direct Orange 61 by Electro-Fenton Process Using a Carbon Felt Electrode: Application of the Experimental Design Methodology. Journal of Electroanalytical Chemistry, 610, 75-84. http://dx.doi.org/10.1016/j.jelechem.2007.07.004

3.   Kesraoui, A., Oturan, N., Bellakhal, N., Dachraoui, M. and Oturan, M.A. (2008) Experimental Design Methodology Applied

to Electro-Fenton Treatment for Degradation of Herbicide Chlortoluron. Applied Catalysis B: Environmental, 78, 334-341. http://dx.doi.org/10.1016/j.apcatb.2007.09.032

4.   Lynch, B.S., Delzell, E.S. and Bechtel, D.H. (2002) Toxicology Review and Risk Assessment of Resorcinol: Thyroid Effects. Regulatory Toxicology and Pharmacology, 36, 198-210. http://dx.doi.org/10.1006/rtph.2002.1585

5.   Iurascu, B., Siminiceanu, I., Vione, D., Vicente, M.A. and Gil, A. (2009) Phenol Degradation in Water through a Heterogeneous Photo-Fenton Process Catalyzed by Fe-Treated Laponite. Water Research, 43, 1313-1322.http://dx.doi.org/10.1016/j.watres.2008.12.032

6.   Kesraoui, A., Oturan, M.A., Oturan, N., Bellakhal, N. and Dachraoui, M. (2010) Treatment of an Aqueous Pesticides Mixture Solution by Direct and Indirect Electrochemical Advanced Oxidation Processes. International Journal of Environmental Analytical Chemistry, 90, 1029-1397.

7.   Sanz, J., Lombrana, J.I., Deluis, A.M., Ortueta, M. and Varona, F. (2003) Microwave and Fenton's Reagent Oxidation of Wastewater. Environmental Chemistry Letters, 1, 45-50.http://dx.doi.org/10.1007/s10311-002-0007-2

8.   Chamarro, E., Marco, A. and Esplugas, S. (2001) Use of Fenton Reagent to Improve Organic Chemical Biodegradability. Water Research, 35, 1047-1050.http://dx.doi.org/10.1016/S0043-1354(00)00342-0

9.   Hoigné, J. (1998) Chemistry of Aqueous Ozone and Transformation of Pollutants by Ozonation and Advanced Oxidation Processes. The Handbook of Environmental Chemistry, Quality and Treatment of Drinking Water II (J. Hrubec, Ed.), Springer, Berlin, 5, 83-141.

10.   Aleboyeh, A., Moussa, Y. and Aleboyeh, H. (2005) The Effect of Operational Parameters on UV/H_2O_2 Decolourisation of Acid Blue 74. Dyes and Pigments, 66, 129-134.http://dx.doi.org/10.1016/j.dyepig.2004.09.008

11.   Daneshvar, N., Rabbani, M., Modirshahla, N. and Behnajady, M.A. (2005) Photooxidative Degradation of Acid Red 27 in a Tubular Continuous-Flow Photoreactor: Influence of

Operational Parameters and Mineralization Products. Journal of Hazardous Materials, 118, 155-160. http://dx.doi.org/10.1016/j.jhazmat.2004.10.007

12. Liu, P., Li, C., Zhao, Z., Lu, G., Cui, H. and Zhang, W. (2013) Induced Effects of Advanced Oxidation Processes. Scientific Reports, 4, 1-4.

13. Herrmann, J.M., Guillard, C., Arguello, M., Agüera, A., Tejedor, A., Piedra, L. and Alba, A.F. (1999) Photocatalytic Degradation of Pesticide Pirimiphos-Methyl: Determination of the Reaction Pathway and Identification of Intermediate Products by Various Analytical Methods. Catalysis Today, 54, 353-367. http://dx.doi.org/10.1016/S0920-5861(99)00196-0

14. Lhomme, L., Brosillon, S., Wolbert, D. and Dussaud, J. (2005) Photocatalytic Degradation of a Phenylurea, Chlortoluron, in Water Using an Industrial Titanium Dioxide Coated Media. Applied Catalysis B: Environmental, 61, 227-235.http://dx.doi.org/10.1016/j.apcatb.2005.06.002

15. Comninellis, C. and Pulgarin, C. (1991) Anodic Oxidation of Phenol for Waste Water Treatment. Journal of Applied Electrochemistry, 21, 703-708.http://dx.doi.org/10.1007/BF01034049

16. Boye, B., Dieng, M.M. and Brillas, E. (2002) Degradation of Herbicide 4-Chlorophenoxyacetic Acid by Advanced Electrochemical Oxidation Methods. Environmental Science and Technology, 36, 3030-3035.http://dx.doi.org/10.1021/es0103391

17. Hanna, K., Chiron, S. and Oturan, M. (2005) Coupling Enhanced Water Solubilization with Cyclodextrin to Indirect Electrochemical Treatment for Pentachlorophenol Contaminated Soil Remediation. Water Research, 39, 2763-2773.http://dx.doi.org/10.1016/j.watres.2005.04.057

18. Bellakhal, N., Dachraoui, M., Oturan, N. and Oturan, M.A. (2006) Degradation of Tartrazine in Water by Electro- Fenton Process. Journal de la société chimique de Tunisie, 8, 223-228.

19. Hammami, S., Bellakhal, N., Oturan, N., Oturan, M.A. and Dachraoui, M. (2008) Degradation of Acid Orange 7 by Electrochemically Generated OH⁻ Radicals in Acidic Aqueous Medium Using a Boron-Doped Diamond or Platinum Anode:

A Mechanistic Study. Chemosphere, 73, 678-684.http://dx.doi.org/10.1016/j.chemosphere.2008.07.010

20. Vinodgopal, K. and Peller, J. (2013) Hydroxyl Radical-Mediated Advanced Oxidation Processes for Textile Dyes: A Comparison of the Radiolytic and Sonolytic Degradation of the Monoazo Dye Acid Orange 7. Journal of Agricultural and Food Chemistry, 29, 307-316.

21. Nam, K., Rodriguez, W. and Kukor, J.J. (2001) Enhanced Degradation of Polycyclic Aromatic Hydrocarbons by Biodegradation Combined with a Modified Fenton Reaction. Chemosphere, 45, 11-20. http://dx.doi.org/10.1016/S0045-6535(01)00051-0

22. Luo, M., Bowden, D. and Brimblecombe, P. (2009) Catalytic Property of Fe-Al Pillared Clay for Fenton Oxidation of Phenol by H_2O_2. Applied Catalysis B: Environmental, 85, 201-206.http://dx.doi.org/10.1016/j.apcatb.2008.07.013

23. Gonçalves, D.B., Teixeira, J.A., Bazzolli, D.M., Queiroz, M.V. and Fernandes, E. (2012) Use of Response Surface Methodology to Optimize Production of Pectinases by Recombinant Penicillium griseoroseum T20. Biocatalysis and Agricultural Biotechnology, 1, 140-146. http://dx.doi.org/10.1016/j.bcab.2011.09.002

24. Khataee, A.R., Zarei, M. and Moradkhannejhad, L. (2010) Application of Response Surface Methodology for Optimization of Azo Dye Removal by Oxalate Catalyzed Photoelectro-Fenton Process Using Carbon Nanotube-PTFE Cathode. Desalination, 258, 112-119. http://dx.doi.org/10.1016/j.desal.2010.03.028

25. Ravikumar, K., Ramalingam, S., Krishnan, S. and Balu, K. (2006) Application of Response Surface Methodology to Optimize the Process Variables for Reactive Red and Acid Brown Dye Removal Using a Novel Adsorbent. Dyes and Pigments, 70, 18-26.http://dx.doi.org/10.1016/j.dyepig.2005.02.004

26. Sayon, E. (2006) Ultrasound-Assisted Preparation of Active Carbon from Alkaline Impregnated Hazelnut Shell: An Optimization Study on Removal of Cu^{2+} from Aqueous Solution. Chemical Engineering Journal, 115, 213-218.http://dx.doi.org/10.1016/j.cej.2005.09.024

27. Zhao, D., Ding, C., Wu, C. and Xu, X. (2012) Kinetics of Ultrasound-Enhanced Oxidation of p-Nitrophenol by Fenton's

Reagent. Energy Procedia: 2012 International Conference on Future Energy, Environment, and Materials, 16, 146-150.

28. Hamzaoui, A., Jamoussi, B. and M'nif, A. (2008) Lithium Recovery from Highly Concentrated Solutions: Response Surface Methodology (RSM) Process Parameters Optimization. Hydrometallurgy, 90, 1-7.http://dx.doi.org/10.1016/j.hydromet.2007.09.005

29. Missaoui, I., Sayedi, L., Jamoussi, B. and Ben Hassine, B. (2009) Response Surface Optimization for Determination of Volatile Organic Compounds in Water Samples by Headspace-Gas Chromatography-Mass Spectrometry Method. Journal of Chromatographic Science, 47, 257-262. http://dx.doi.org/10.1093/chromsci/47.4.257

30. Myer, R. and Montgomery, D.C. (2002) Response Surface Methodology. John Wiley and Sons Inc., New York.

31. Luster, J., Kalbitz, K., Lennartz, B. and Rinklebe, J. (2014) Properties, Processes and Ecological Functions of Floodplain, Peatland, and Paddy Soils. Geoderma, 228-229, 1-4.http://dx.doi.org/10.1016/j.geoderma.2014.04.010

32. Baccar, A., Batis, N. and Ghorbel, A. (2005) Facts of Parameters and Protocol Preparation on Structural and Textural Properties of Iron (III) Intercalated Clay. Journal de la société chimique de Tunisie, 7, 173-186.

Chapter 10

Studies in Molecular Weight Determination of Cottonseed and Melon Seed Oils Based Biopolymers

Ibanga O. Isaac and Edet W. Nsi

Department of Chemistry, Akwa Ibom State University, Uyo, Nigeria

ABSTRACT

Six grades of biopolymers formulated to have oil content of 40% (M_1), 50% (M_2), and 60% (M_3) melon seed oil (MESO) and 40% (C_1), 50% (C_2), and 60% (C_3) cottonseed oil (COSO) respectively, were prepared with phthalic anhydride, and glycerol using alcoholysis-polycondensation process. The extend of polycondensation was monitored by determining the acid value of aliquots of the reaction mixture at various intervals of time. Molecular weight averages and polydispersity index (PDI) of the finished alkyds were determined by Rast method and end-group analysis. Molecular weight averages and PDI vary with differences in oil length of the alkyds, with samples M_2 and C_2 respectively exhibiting the highest PDI. Molecular weight average obtained from end-group analysis and those determined by Rast method in brackets are 1338.92 (597.00), 982.33 (696.25), 1316.09 (754.03), and 1160.57 (448.13), 765.96 (583.57), 1049.92 (696.25) for samples M_1, M_2, M_3 and C_1, C_2, C_3 respectively. Number molecular weight averages calculated from end-group analysis are larger than those obtained by Rast method for both MESO and COSO alkyds and seem to grossly overestimate their molecular weights. The mode of variation of these properties indicates that the synthesis of MESO and COSO alkyds are complex. Correlation of PDI with the quality of the finished alkyds shows that the higher the PDI value the better the quality of the alkyd. Performance properties such as rate of drying, film hardness and resistance to chemicals were optimum at 50% oil length for both triglyceride oil alkyds.

INTRODUCTION

Alkyd resins have been identified as one of the biopolymers synthesised from triglyceride oils [1] . It is a key ingredient of all surface-coating products like paints, primer, adhesives, printing ink [2] and varnishes. They are tough resinous materials prepared via an esterification reaction among polybasic acids, polyols and monoacids (commonly fatty acids from vegetable oils or fats) [3] .

Alkyd resins have been recognized as complex systems because of the variety of ingredients used in their preparation and the possibility of formation of three-dimensional network [4] [5] . Alkyd resin usually consists of mixtures of low, medium and high molecular weight species. Hence, its properties depend on the average molecular weight and molecular weight distribution. It is well known that the end-use properties of alkyd resin as coatings become optimum at a region along the extent of reaction co-ordinate when appreciable larger molecules begin to form, i.e., cross-linking of molecular chains [5] [6] . Thus, the physical, chemical and mechanical properties of alkyd resins like other polymers are dependent on their relative sizes or molecular weight [5] [7] [8] .

Generally, properties of polymers are determined by the molecular properties—molecular structure, polarity and flexibility of the polymer chains. For instance, rate of oxidation of film deposited, and its toughness and resistance to degradation, have been found to be related to the molecular weight of the alkyd [7] [9] . Similarly, the stability of alkyds on storage and their solubility in conventional hydrocarbon solvents, for example, white spirit, commonly employed in modifying their viscosity, are both dependent on their average molecular weights [8] . Therefore, knowledge of the average molecular weight and molecular weight distribution of the finished alkyds is of utmost importance in their practical application [8] and as a means of evaluating their performance as binders in surface coatings.

However, in processing alkyds for optimum coating performance, it is required that the molecular weights are not too large; otherwise, the alkyd may convert into an intractable gel and become out of control or too low such that its coating performance is impaired [8] . Thus, alkyd resins of appreciable molecular sizes that would ensure trouble-free processing, exhibit stability on storage and perform well during application are desirable for the alkyd chemists and technologists.

In our previous studies [10] [11] , it has been discovered that melon seed and cottonseed oils are potential raw material for substituting imported drying oils for use in the Nigerian surface coating industry. Hence, the need for further research involving the determination of molecular weight of these products uses different methods in order to ascertain the method suitable for routine analysis of the alkyds.

The use of cryoscopic or Rast method for determination of the weight-average molecular weight, $\overline{M}_w$, number-average molecular weight, $\overline{M}_n$, and polydispersity of alkyd resins have been reported [4] [8] [12] - [14] . Absolute methods of molecular weight determination such as light scattering and osmometry require sophisticated and expensive equipment, often; experiments take long time to complete. Many industries cannot afford such sophisticated equipment and lack personnel to handle it. Thus, such methods are unsuitable for routine analysis required in the surface coating industry, since empirical rather than absolute molecular weights are needed. This research is therefore aimed at comparing the average molecular weights of cottonseed and melon seed oils alkyd resins obtained by end-group analysis and Rast method, in order to ascertain the method suitable for routine analysis of molecular weights of these alkyd resins.

MATERIALS AND METHODS

Materials

Triglyceride oils used for alkyd resins preparation were locally sourced. COSO was purchased at Sabongari market, Kano, while melon seeds were purchased at Akpan Andem market, Uyo, processed and the oil soxhlet extracted using the method described in our previous work [11] . Technical grade phthalic anhydride, glycerol, xylene and lead (II) oxide were obtained from commercial sources and used without further purification in the preparation of COSO and MESO alkyd resins. A micro-analytical reagent (MAR) camphor obtained from commercial source was used in the determination of melting points of unfractionated alkyd samples.

Preparation of Alkyd Resins

Six grades of alkyd resins formulated to have oil content of 40% (M_1), 50% (M_2), and 60% (M_3) MESO and 40% (C_1), 50% (C_2), and 60% (C_3) COSO respectively, which were prepared with phthalic anhydride, glycerol and xylene in the previous study [10] , were used in this experiment. Aliquots of the reaction mixture were withdrawn at 20 min intervals and the acid value determined by titrimetric method [15] . The recipe for the formulation of short, medium and long oil alkyds of MESO and COSO respectively reported elsewhere [16] .

Determination of Molecular Weights of Alkyd Samples

Rast method and end-group analysis were employed in the determination of the average molecular weights of the alkyd samples.

Cryoscopy (Rast Method)

Number-average molecular weights ($\overline{M}_{det}$) of the finished alkyds were determined by the cryoscopic method employing the freezing-point depression of camphor [14] . Alkyd sample (0.05 g) was weighed into a test tube, then, 0.5 g of pure camphor (MAR) was weighed into the test tube. The test tube was stoppered loosely with a cotton wool and the content was melted by placing it in an oil bath, previously heated to about 140°C, for about 1 min. It was allowed to cool down and the content was transferred to a clean watch glass and powdered. Some of the powder was introduced into a thin capillary tube of which one end was carefully sealed in a flame and its melting point determined. The $(\overline{M}_{det})$ was computed from the expression in Equation (1).

$$\overline{M}_{det} = \frac{k \times y \times 1000}{(T_2 - T_1) \times Z}$$

(1)

where y is the weight (g) of alkyd sample, Z is the weight (g) of pure camphor, T_1 is the melting point of the mixture of camphor and alkyd, T_2 is melting point of pure camphor and k is the molecular depression constant of camphor and equals 39.7 [14] .

End-Group Analysis

The extent of the reaction (P_n), and number-average degree of polymerization $(\overline{X}_n)$ calculated in our previous work [1] were used in the determination of weight-average molecular weight $(\overline{M}_w)$, and number-average molecular weight $(\overline{M}_n)$, as well as PDI based on Equations (2), (3) and (5) respectively.

$$\overline{M}_w = \frac{w}{(k - P_n)e_A}$$

(2)

where w is the total weight of all the ingredients charged, k, the ratio of the total moles of all ingredients charged to total equivalents of the acid (M_o / e_A); and e_A, the total equivalent of the acid. These parameters were calculated in our previous studies [16] .

$$\bar{M}_n = \frac{M_o}{(1 - P_n)}$$

(3)

where M_o is the average molecular weight of the repeating unit of the alkyd and it is expressed as follows:

$$M_o = \frac{Q(MW)_1 + P(MW)_2 + G(MW)_3}{Q + P + G}$$

(4)

where Q, P and G are respectively the number of moles of the oil, diacid and glycerol used in the formulation [16] and $MW_1 = 265.95$, $MW_2 = 132.00$, and $MW_3 = 89.60$ are the approximate molecular weights of oil, diacid and glycerol, respectively, corrected for the possible loss of end groups [5] [7] [8] .

$$\bar{M}_w / \bar{M}_n = PDI$$

(5)

RESULTS AND DISCUSSION

Molecular Weight Characterization of MESO and COSO Alkyd Resins

It is observed from Figure 1 and Table 1 that during the early stages (up to 60 min) of MESO alkyd synthesis, there seems to be no substantial increase in the molecular sizes of the reaction mixture. However, an appreciable increase was observed as the reaction progressed. Similar observation was made for COSO alkyd samples (C_1)- (C_3), such that during the early stages of reaction (up to 80 min), no substantial increase in the molecular weights were observed (Figure 2 and Table 1).

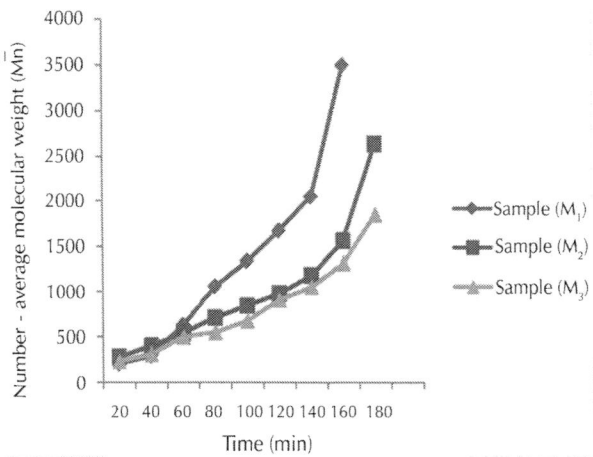

Figure 1: Plot of number-average molecular weight of melon seed oil alkyd sample M_1, M_2 and M_3 against reaction time.

Table 1: Weight-average molecular weight ($\overline{M_w}$) of melon seed and cotton-seed oil-modified alkyd resins at various stages of reaction

Time	Sample (M1)	Sample (M2)	Sample (M3)	Sample (C1)	Sample (C2)	Sample (C3)
20	232.38	335.27	252.39	197.350	235.63	251.47
40	340.37	539.76	344.94	250.920	320.72	347.71
60	891.80	872.093	591.90	442.217	670.54	586.63
80	2112.08	1434.03	659.30	1148.545	826.45	644.88
100	3597.12	2215.66	854.96	1872.659	1390.18	860.09
120	7727.98	3588.52	1275.92	2551.020	1716.25	1232.54
140	51724.14	10416.67	1576.27	9554.140	5474.45	1564.13

M_1—40% MESO alkyd sample; M_2—50% MESO alkyd sample; M_3—60% MESO alkyd sample; C_1—40% COSO alkyd sample; C_2—50% COSO alkyd sample; C_3—60% COSO alkyd sample.

However, an appreciable increase was observed, especially for sample (M_1) after about 120 min followed by sample (M_2) and finally M_3 having the least values for MESO alkyd samples. Similarly, the molecular sizes of COSO alkyd reaction mixture increases from C_1 through C_2 with C_3 having the least value after 180 min of reaction (Figure 2). This observation may be because the two oils are semi-drying oil [10] [11] . Further more, the shapes of these plots depict the complexity in the changes in the molecular sizes of the reaction mixture. The low molecular weight of the alkyds observed at the beginning of the reaction was attributed to the low rate of polymerization. Increase in the polymerization rate resulted in the corresponding increase in molecular weight. This trend continued until the gelation point, at which structural changes began to occur in the polymer size of the molecules in solution. This observation is in agreement with literature reports for castor oil and rubber seed oil alkyds [5] [8] [17] . It is also evident from Figure 1 and Figure 2 and Table 1 that the number—average molecular weight as well as weight—average molecular weight of the in-process samples decreased with an increase in the oil lengths as also reported in literature for rubber seed oil and castor oil alkyd resins [8] [17] . Nagata [6] suggested that an increase in oil length increases the amount of fatty acid available for reaction, which in turn results in a higher chance of termination of chain growth, resulting in lower molecular weight polymers. The above statement is applicable if fatty acid is used. On the other hand, when triglyceride oil is used as in this research, the termination might be due to the possibility of the presence of traces of diglycerides in the reaction mixture, if the triglycerides were not completely converted to monoglycerides during the alcoholysis stage of the reaction. The presence of diglycerides can terminates the polymerization and hence low chain length. The trend in the variations of molecular weights of MESO alkyds were as follows: 40% crude melon seed oil alkyd sample (M_1) greater than 50% crude melon seed oil alkyd sample (M_2) and 60% crude melon seed oil alkyd sample (M_3). Those of COSO alkyds were similar to its MESO counterparts: C_1 greater than C_2 and C_3.

The weight—average molecular weight, ($\overline{M}_w$), the number—average molecular weight $\overline{M}_n$ and polydispersity of MESO and COSO alkyd resins calculated at the region where deviation from linearity was observed in Figure 1 and Figure 2 as well as average molecular weight determined using cryoscopic method, $\overline{M}_{det}$, are presented in Table 2. On the other hand, Figure 3 and Figure 4 compare the $\overline{M}_{det}$ and $\overline{M}_n$ of MESO and COSO alkyd resins respectively.

From Table 2 and Figure 3, it is observed that $(\overline{M}_w)$, decreases from 3597.12 for sample M_1, 3588.52 for sample M_2 and 1576.27 for alkyd sample M_3. $\overline{M}_n$ values range from 982.33 for sample M_2, 1316.09 for sample M_3 and 1338.92 for sample M_1. Values of $\overline{M}_{det}$ on the other hand, increases from 597.00 for sample M_1, 696.25 for sample M_2 and 754.03 for sample M_3.

The $\overline{M}_n$ obtained at the region where there was deviation from linearity ranged from 765.96 for alkyd sample C_2, 1049.92 for alkyd sample C_3 and 1160.57 for cottonseed oil alkyd sample C_1. Also, values of $(\overline{M}_w)$ decreases from 2551.02 for sample C_1, 1716.25 for sample C_2 and 1564.13 for sample C_3. On the other hand,

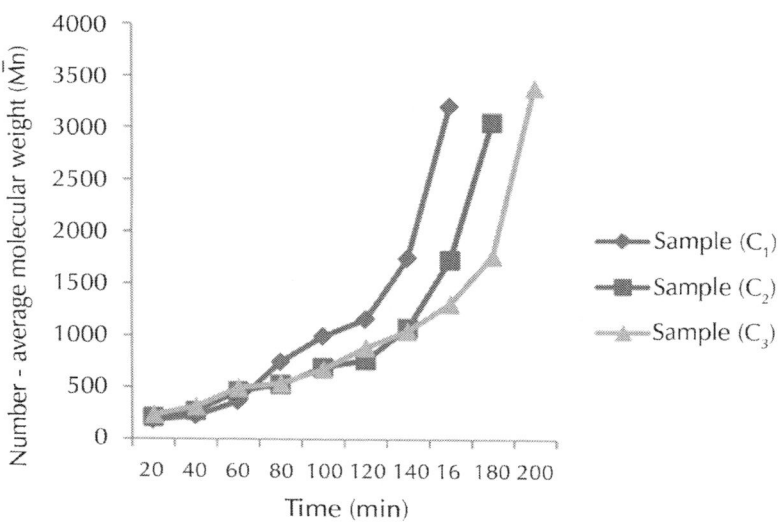

Figure 2: Plot of number-average molecular weight of cottonseed oil alkyd sample C_1, C_2 and C_3 against reaction time.

Table 2: Molecular weight averages of MESO and COSO alkyds obtained by cryoscopy ($\overline{M}_{det}$), and those obtained by end-group analysis (at the point of deviation from linearity from Figure 1, Figure 2) (($\overline{M}_w$) and $\overline{M}_n$) and polydispersity index (PDI)

Alkyd Sample	Average molecular weight			PDI
	$\overline{M}_{det}$	($\overline{M}_w$)	$\overline{M}_n$	($\overline{M}_w$ / $\overline{M}_n$)
M1 40% MESO alkyd sample	597.00	3597.12	1338.92	2.69
M2 50% MESO alkyd sample	696.25	3588.52	982.33	3.65
M3 60% MESO alkyd sample	754.03	1576.27	1316.09	1.19
C1 40% COSO alkyd sample	448.13	2551.02	1160.57	2.19
C2 50% COSO alkyd sample	583.57	1716.25	765.96	2.24
C3 60% COSO alkyd sample	696.25	1564.13	1049.92	1.49

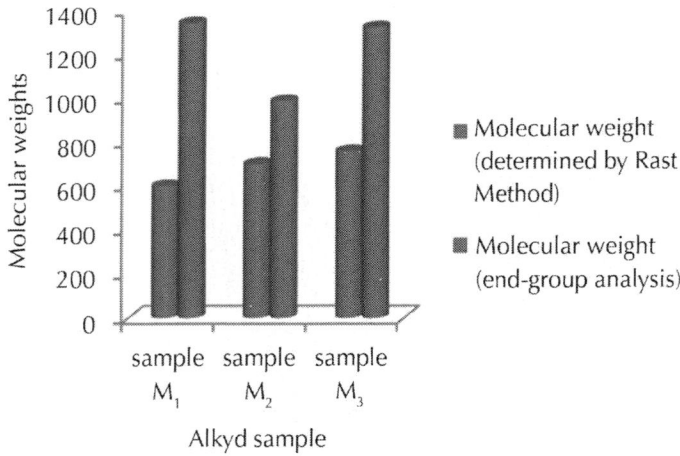

Figure 3: Molecular weights ($\overline{M}_{det}$ and $\overline{M}_n$) for melon seed oil alkyd sample (M_1), (M_2) and (M_3).

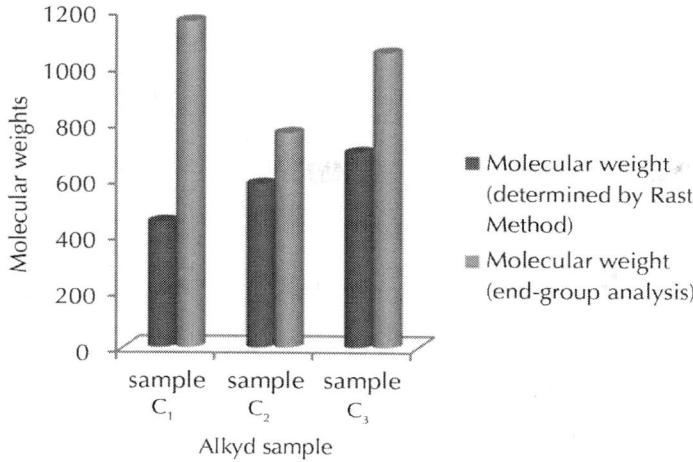

Figure 4: Molecular weights ($\overline{M}_{det}$ and $\overline{M}_n$) for cottonseed oil alkyd samples (C_1), (C_2) and (C_3).

$\overline{M}_{det}$ value for cottonseed oil alkyd samples increases from 448.13 for sample C_1, 583.57 for sample C_2 and 696.25 for sample C_3 (Table 2 and Figure 4). These results is in collaboration with those of crude melon seed oil alkyds counterpart above and are comparable with those reported in the literature for rubber seed oil alkyds resins [4] .

It differs from those reported by [8] in which the $\overline{M}_n$, ($\overline{M}_w$) and PDI in parentheses for short, medium and long oil alkyd samples of rubber seed oil obtained by gel permeation chromatography were 3234, 6186 (1.9); 1379, 2147 (1.56); and 3304, 8406 (2.56) respectively. The medium oil had the least PDI value, while in this research; the medium oil had the highest PDI value. This could be attributed to the difference in method used for obtaining $\overline{M}_n$ and ($\overline{M}_w$) values, which in this case is end-group analysis. The PDI values as observed in Table 3 vary with reaction time with short and medium oil alkyd of both MESO and COSO alkyd samples being higher than that of long oil alkyds of the two oil samples. Hence, the PDI obtained by end-group analysis increases with increase degree of polymerization. The unusual high value of PDI of 23.63 for M_1 compared to it C_1 counterpart of 5.48 (Table 3), could be due the fact that crude and not refined MESO was used for preparation alkyd sample M_1, while refined COSO was used for preparation of alkyd sample C_1.

These molecular weight averages and PDI show that the reaction mixture is constituted of low molecular weight species. This could be due to the possibility of the presence of diglycerides, which terminates the polymerization, and hence low chain length. The PDI clearly indicate that the size distribution is broad. However, 50% MESO alkyd sample M_1 and 50% COSO alkyd sample C_1 with the highest PDI value of 3.65 2.24 respectively can be considered as having the broadest molecular weight distribution.

The results on Table 2 as well as Figure 3 and Figure 4 also show that $\overline{M}_n$ calculated from end-group analysis are larger than $\overline{M}_{det}$ determined from Rast method for both melon seed oil and cottonseed oil alkyds.

Table 3: Polydispersity index of MESO and COSO alkyd samples

Time	Sample (M1)	Sample (M2)	Sample (M3)	Sample (C1)	Sample (C2)	Sample (C3)
20	1.11	1.20	1.08	1.10	1.12	1.08
40	1.16	1.35	1.12	1.12	1.19	1.11
60	1.42	1.60	1.19	1.22	1.45	1.19
80	1.99	2.03	1.21	1.54	1.57	1.20
100	2.69	2.62	1.27	1.88	1.20	1.27
120	4.62	3.65	1.40	2.20	2.24	1.39
140	23.63	8.81	1.49	5.48	5.08	1.49

The discrepancy between $\overline{M}_{det}$ and $\overline{M}_n$ may be attributed to the assumptions made in the end-group analysis to the effect that the reactivity of the polymer chains is independent of its size, that the functional groups of the same kind are equally reactive, that intramolecular reactions are absent, and that only interesterification reaction occurs during polyesterification [4] [8] [13] . Actually, occurrence of intramolecular condensation or cyclization leading to ring formation during polyesterification has been reported [12] . Since an intramolecular reaction could occur during alkyd preparation leading to a decrease in concentration of carboxyl groups without a corresponding increase in chain length, there will be an increase in extent of reaction and consequently average degree of polymerization calculated thereof. It is not surprising therefore, that in the present study, $\overline{M}_n$, calculated from end ? group analysis is larger than those determined by Rast method($\overline{M}_{det}$). Hence, end-group analysis seems to grossly overestimate the $\overline{M}_n$ of the alkyd resins. Indeed, similar discrepancies between $\overline{M}_{det}$ and $\overline{M}_n$ have been reported [5] [8] [18] .

Previous studies on molecular weight determination of some alkyds indicated that they are constituted of species of relatively low molecular weights of approximately 500 [12] [18] , although molecular weights of between 1170 and 1250 have been reported [6] . It is observed that both $\overline{M}_{det}$ and $\overline{M}_n$ for all the alkyd samples studied fall within the range reported by Aigbodion and Pillai [8] and Flory [12] .

Correlation of the performance characteristics of the melon seed and cottonseed oils alkyds obtained in our previous studies [10] [11] [19] with their number-average molecular weights and PDI shows that the lower the polydispersity index of the alkyds are the less is their performance as binders. Thus, 50% oil length alkyds (M_2 and C_2), which has the highest PDI, exhibited the best characteristics such as excellent adhesion, good hardness, fast drying time and resistance to chemicals followed by 40% oil length alkyds (M_1 and C_1). 60% oil length alkyds (M_3 and C_3) with the least PDI had the least performance characteristics.

CONCLUSIONS

In conclusion, it can be inferred from these results that reactions leading to the formation of melon seed and cottonseed oils alkyd resins are complex. The PDI of MESO and COSO alkyd resins is an important parameter, which determines their performance as binders in surface coating products. Average molecular weights determined by Rast method increased with increase in oil length of the alkyds and were less than those obtained from end-group analysis. It is found in this study that Rast method is more reasonable than end-group analysis for routine determination of molecular weights of alkyd resins.

REFERENCES

1. Isaac, I.O. and Ekpa, O.D. (2014) Comparative Study on the Kinetics of the Preparation of Melon Seed and Cottonseed Oils Based Biopolymers. American Journal of Polymer Science, 4, 7-15.

2. Uzoh, C.F., Onukwuli, O.D., Odera, R.S. and Ofochebe, S. (2013) Optimization of Polyesterification Process for Production of Palm Oil Modified Alkyd Resin Using Response Surface Methodology. Journal of Environmental Chemical Engineering, 1, 777-785. http://dx.doi.org/10.1016/j.jece.2013.07.021

3. Ling, J.S., Mohammed, I.A., Ghazali, A. and Khairuddean, M. (2014) Novel Poly (Alkyd-Urethane)s from Vegetable Oils: Synthesis and Properties. Industrial Crops and Products, 52, 74-84. http://dx.doi.org/10.1016/j.indcrop.2013.10.002

4. Aigbodion, A.I. and Okieimen, F.E. (1996) Kinetics of the Preparation of Rubber Seed Oil Alkyds. European Polymer Journal, 32, 1105-1108. http://dx.doi.org/10.1016/0014-3057(96)00053-5

5. Okieimen, F.E. and Aigbodion, A.I. (1997) Studies in Molecular Weight Determination of Rubber Seed Oil Alkyds. Industrial Crops and Products, 6, 155-161.http://dx.doi.org/10.1016/S0926-6690(96)00209-9

6. Nagata, T. (1969) Cooking Schedule of Alkyd Resin Preparation—Part II. Effect of Cooking Schedule on Molecular Weight Distribution of Alkyd Resin. Journal of Applied Polymer Science, 13, 2601-2619. http://dx.doi.org/10.1002/app.1969.070131208

7. Bobalek, E.G., Moore, E.R., Levy, S.S. and Lee, C.C. (1964) Some Implications of Gel Point Concept on the Chemistry of Alkyd Resins. Journal of Applied Polymer Science, 8, 625-657. http://dx.doi.org/10.1002/app.1964.070080207

8. Aigbodion, A.I. and Pillai, C.K. (2001) Synthesis and Molecular Weight Characterization of Rubber Seed Oil-Modi- fied Alkyd Resins. Journal of Applied Polymer Science, 79, 2431-2438. http://dx.doi.org/10.1002/1097-4628(20010328)79:13<2431::AID-APP1050>3.0.CO;2-A

9.   Flory, P.J. (1953) Principles of Polymer Chemistry. Cornell, Ithaca, New York.

10.   Isaac, I.O. and Ekpa, O.D. (2013) Fatty Acid Composition of Cottonseed Oil and Its Application in Production and Evaluation of Biopolymers. American Journal of Polymer Science, 3, 13-22.

11.   Ekpa, O.D. and Isaac, I.O. (2013) Fatty Acid Composition of Melon (Colocynthis vulgaris Shrad) Seed Oil and Its Application in Synthesis and Evaluation of Alkyd Resins. IOSR Journal of Applied Chemistry, 4, 30-41. http://dx.doi.org/10.9790/5736-0443041

12.   Flory, P.J. (1941) Molecular Size Distribution 3-Dimensional Polymers I: Gelation. Journal American Chemical Society, 63, 3083-3090. http://dx.doi.org/10.1021/ja01856a061

13.   Stockmayer, W.H. (1952) Molecular Distribution in Condensation Polymerization. Journal of Polymer Science, 9, 69- 71. http://dx.doi.org/10.1002/pol.1952.120090106

14.   Furniss, B.S., Hannaford, A.J., Rogers, V., Smith, P.W.G. and Tatchel, A.R. (1978) Vogel's Textbook of Practical Organic Chemistry. 4th Edition, ELBS & Longman, London.

15.   Ekpa, O.D. and Isaac, I.O. (2009) Kinetic Studies on Polyesterification of Unsaturated Oils and Diacids in the Alcoholoysis Process. Research Journal of Applied Science, 4, 125-128.

16.   Isaac, I.O. and Ekpa, O.D. (2013) Comparative Study on Solution Viscosity Properties of Cottonseed and Melon Seed Oils Based Biopolymers. American Journal of Polymer Science, 3, 23-34.

17.   Onukwli, O.D. and Igbokwe, P.K. (2008) Production and Characterization of Castor Oil-Modified Alkyd Resins. Journal of Engineering & Applied Science, 3, 161-165.

18.   Kienle, R.H., Van der Meulen, P.A. and Petke, F.E. (1939) The Polyhydric Alcohol-Polybasic Acid Reaction III: Further Studies of the Glycerol-Phthalic Acid Reaction. Journal of American Chemical Society, 61, 2258-2268. http://dx.doi.org/10.1021/ja01878a001

19.   Isaac, I.O. and Nsi, E.W. (2013) Influence of Polybasic Acid Type on the Physicochemical and Viscosity Properties of Cottonseed Oil Alkyd Resins. The International Journal of Engineering and Science, 2, 1-14.

11

Rheological Characterization of a Mixed Fruit/Vegetable Puree Feedstock for Hydrogen Production by Dark Fermentation

Jacob Gomez-Romero[1], Inés Garcia-Peña[1], Jorge Ramirez-Muñoz[2], and Luis G. Torres[1*]

[1]Bioprocess and Bioengineering Departments, UPIBI-Instituto Politecnico Nacional, Mexico City, Mexico

[2]Energy Department, Universidad Autónoma Metropolitana-Azcapotzalco, Mexico City, Mexico

ABSTRACT

Bio-hydrogen (Bio-H_2) production from the organic fraction of solid waste, as fruit and vegetable wastes, constitutes an interesting and feasible technology to obtain clean energy. In spite of the feasibility to produce Bio-H$_2$ from fruit/vegetable wastes (FVW), data about its rheological characterization are scarce. This information is useful to establish the hydrodynamic behavior, which controls the overall mixing process when the feedstock for Bio-H$_2$ production process is a mixture of FVW. In this work, the rheological behavior of a vegetable/fruit waste mixture was characterized. The effect of the solids content (%, w/w), temperature, time (tyxotropy effects) and shear rate over the apparent viscosity of the mixture was evaluated. Most of the mixtures showed non-Newtonian behavior. The curves are typical rheofluidizing fluids. The rheological curves were different at increasing solids contents (80%, 60%, 40% and 30%), independent from the temperature. Rheological data were fitted to the power law model. Correlation factors R2 for the different mixtures were 0.991 - 0.995 for 80%, 0.961 - 0.986 for 60%, 0.890 - 0.925 for 40%. In the case of 30% of solids, the R2 value was not acceptable, and it was also found that this mixture was very near to the Newtonian behavior. Calculated activation energies (Ea) values were 15.98, 14.89 and 20.96 kJ/mol for the 80%, 60%, 40% mixtures, respectively. FVW purees rheological behavior was well characterized by Carbopol solutions at given concentrations and pH values. This fluid can be used as a model for other studies, e.g. LDA (Laser Doppler Anemometry) and PIV (Particle Image Velocimetry).

INTRODUCTION

Bio-Hydrogen Production from Vegetable and Fruit Wastes FVW

Recently, the research activity has been focused on the production of alternative energy sources from organic matter [1]. Additionally, FVW are produced in large quantities in markets in many large cities [2-5]. Mexico City with a population of more than 20 million produces a huge amount of solid waste greater than 12,000 tons per day, which has to be disposed. On the other hand, the enough food supply for this population is provided by many markets located at different regions of the city. The largest market is the Central de Abastos (CEDA), which commercializes 24,000 tons of food products and produces 895 tons of organic solid waste each day. The waste produced by CEDA is conveniently separated at the source, and is an ideal candidate for energy production using anaerobic processes [6]. Vegetable and fruit waste produced in markets is considered as a potential source to generate energy due to its higher organic composition and easily biodegradable nature. The application of an anaerobic digestion process for simultaneous waste treatment and renewable energy production from the organic fraction of these residues could therefore be of great interest [5].

Bio-hydrogen (Bio-H$_2$) production from the organic fraction of solid waste, as fruit and vegetable wastes, constitutes an interesting and feasible technology to obtain clean energy. Vegetable-based waste contains a high amount of carbohydrate [7]. The feasibility of Bio-H$_2$ production from the carbohydrate rich wastes, such as food processing wastewater [8], mixed fruit peel-waste [9] and cheese processing wastewater [10] by dark-fermentation process has been reported by some authors. Until now the dark fermentation process for Bio-H$_2$ production has been utilizing different organic market waste or fruit vegetables wastes mixing such as tomato, potato, carrot, cabbage, brinjal, beet-root, okra and coccinia [7,11], lettuce, fennels, lemons, tomatoes, plums apples, strawberries pears and peaches [12] and fruit peel waste [9].

In spite of the feasibility to produce Bio-H$_2$ from FVW [7,13], information about its rheological characterization is very scarce and there is still a need to study the hydrodynamic performance (micro/macromixing) inside a pilot or industrial CSTR that may cause fruit/vegetables waste mixing problems in the process.

Ruggeri and Tommasi [12] conclude that one of the most important parameters that affect Bio-H$_2$production is rheological behavior of the broth and mixing when organic refuse is utilized. They mentioned that the rheological behavior and mixing are ones of the most important problems that need to be solved. For instance, for a highly shear-thinning fluid, the apparent viscosity near the impeller is low, and near the wall, it is very high. Thus, the mixing process can be efficient near the impeller and low near the wall, i.e. the overall mixing process of CSTR's can be drastically affected. Additionally, theoretical and experimental investigations on micro/ macro-mixing aspects are still necessary to scale-up and design a full-scale plan. Therefore, the understanding of the hydrodynamic system is important because this is a critical design parameter in the assessment of the fullscale practical application of fermentation Bio-H$_2$.

Rheological Characterization of Vegetable and Fruits Purees

Some works have reported the rheological characterization of mango and papaya nectar blends[14], blueberry [15,16], zapote [17], raspberry, strawberry, prune and peach [18]. On the other hand, some vegetables or milled seeds have been characterized, such as tomato [19], ginger and chili [20,21], and milled maize and soy in water [22].

As far as we know, there are no previous reports on the characterization of fruit-vegetable waste mixture including pineapple, banana, orange, papaya, tomato and lettuce. This mixture was used as a model substrate for anaerobic digestion process because they are the most sold products commercialized in the CEDA and they are not seasonal products. Some rheological models have been employed for the rheological characterization of fruits/vegetables purees including the Ostwald de Wale power model (various works, including Nindo et al., [16]), the Bingham plastic model, the IPC Paste analysis model, Casson model [14], Herschel-Bulkley model and the Mazhari Berk model [15]. In other works, more complex models, such as the Carrau model, have been reported [19].

In this work, data were fitted only to the power law model [14]:

$$\eta = K \cdot \gamma^{n-1}$$

(1)

where:

is the fluid viscosity in Pa·s;

K is the consistency index, in Pa·s^{n-1};

n is the flow index (undimensional).

Regarding the effect of the temperature over the apparent viscosity of fruit/vegetable puree, the most used equation is the Arrhenius type model and the linear model [17].

The Arrhenius type equation is given by [14]:

$$\eta = \eta_{\infty} \exp\left(Ea/RT\right) \tag{2}$$

where:

η is the apparent viscosity at a fixed shear rate, in Pa·s;

η_{∞} is the viscosity at infinite temperature, in Pa·s;

Ea is the Activation energy in kcal/mol;

R is the universal gas constant (8.314 kJ/kmol·K);

T is the temperature, in K.

In a CFD report, a combined model which uses the power law model and a function with the temperature effect was reported [23,24]:

$$\eta = K\gamma^{n-1}H\left(T\right) \tag{3}$$

Where

$$H\left(T\right) = \left[\alpha\left(1/T - T_0 - 1/T_a - T_0\right)\right] \tag{4}$$

Simulation of Viscous Fluids Using Cellulose-Derivatives

Cellulose-derivatives such as Carbopol have been employed for the simulation of fermentation broths since they are quite soluble, transparent and offer a variety of rheological behavior which depends on either the agent concentration or the pH value. In addition, the use of Carbopol solutions or other yield-stress transparent fluids such as Carboxy Vinyl Polymers (CPV) as model fluids is particularly useful for several visualization techniques in performing mixing experiments, e.g. LDA (Laser Doppler Anemometry), PIV (Particle Image Velocimetry), visualization with dye and flow observation with tracers [25-27]. Galindo and Nienow [28] employed Carbopol solutions to simulate highly-viscous fermentation broths for production of xanthan gum. They studied the mixing of highly viscous (and shear thinning) simulated xanthan fermentation broths with the Lightning A- 315 impeller.

On the other hand, Kelly and Gigas [29] also employed Carbopol solutions (0.1% w/v) to mimic shearrate thinning fluids, for a CFD study with near-axial-flow impellers operating in the transitional regime. Finally, Curran et al. [30] have employed Carbopol solutions as models for yield-stress fluids. They have emphasized that these solutions at a given concentration show different K and n values, depending on the pH value (i.e. from 2.79 to 11.36).

The aim of this work was to characterize the rheological behavior of mixed vegetables/fruits waste mixture which will be used as feedstock for the production of bio-hydrogen by dark fermentation in a 22 L CSTR. The effect of the solids content (%, w/w), the temperature, time (tyxotropyc effects) and the shear rate over the apparent viscosity of the mixture was evaluated using a low-cost viscometer. Finally, the fluid rheology was simulated using Carbopol (a cellulose derivative) solutions, which are clear-translucent viscous fluids, with a rheology behavior dependent from pH values. That fluid will be employed in the real bioreactor in order to carry out mixing experiments (power consumption, PIV, mixing times with dye observation, etc.).

MATERIALS AND METHODS

Puree Elaboration and Characterization

The fruits/vegetables puree was prepared using fruits and vegetables purchased in a local market. All products were mature. Fruits were not pealed and both the skin and the pulp were employed in the puree manufacture. All ingredients were milled separately in a domestic blender during 5 - 10 min adding certain amount of water. At the end, the milled fruits and vegetables were put together, and passed through an stainless steel American mesh 10 (diameter of 2 mm). The process was carried out at environmental conditions. The puree was stored at 4°C, until its dilution at 80%, 60%, 40%, 30% and 20% (w/v) using water again. Volatile fatty acids (VFA) such as lactic acid, acetic acid and ethanol were analyzed by high performance liquid chromatography (HPLC) (Varian ProStar, Model 350), with Rezex ROA organic acid column (Phenomenex®, Torrance, California, USA), equipped with a UV/Vis, RI detector. Solution sulfuric acid (0.005 N) was used as the mobile phase, and separation were carried out at 65°C at a flow rate of 0.6 mL/min and pressure at 39 bar. Previously, the samples were centrifuged at 10,000 rpm (Eppendorf®, Mini Spin) for 10 min, after which the supernatant was filtered (Whatman®, 0.45 µm) to analysis. Total sugars (glucose and fructose) were analyzed by the di-nitro salicylic acid (DNS) method, while soluble protein was determined by the Bradford method with bovine serum albumin as standard. COD was measured using Hach Standard Method (KitHach, 20 a 1500 ppm). The pH, total solids (TS), and volatile suspended solids (VSS), were measured according to standard methods (APHA, 2005). Liquid samples for metal analysis were filtered (Millipore, Sigma-Aldrich, 0.45 µm) and metals analyzed were iron and nickel with an atomic absorption spectroscopy (Perkin Elmer, USA). Carbon, nitrogen, oxygen, phosphorus, sulfur and metals as sodium, magnesium, potassium and calcium were undertaken using Energy-dispersive X-ray spectroscopy technique by elemental analysis (Ray-X Bruker D8 Advance).

Rheological Characterization

Rheological characterization was carried out in a lowcost viscometer (Brookfield Model DV-II + Pro, USA) in the range of 0.667 - 3.33 s^{-1}. For that, different spindles were employed. Measurements were run always by duplicate, once upwards and once downwards, in order to check any tyxotropyc effect. After that, the average viscosity was recorded for every puree concentration, shear rate and temperature.

Measurements were carried out at 20°C, 30°C and 40°C, since the hydrogen fermentation has been developed around 35°C. For that, a hot water bath was employed, and the puree sample was equilibrated at the desired temperature before the beginning of the specific measurement.

Rheological Behavior of the Carbopol Solutions

Carbopol solutions of different concentrations adjusted to various pH values were prepared and characterized as previously explained at 20°C only. Samples were kept at 4°C previous to use.

RESULTS

Purees Characterization

Organic matter, total solids, volatile solids and pH were evaluated for the fruits and vegetables apart, and for the mixture. Results, previously reported by Garcia-Peña et al. [6] are shown atTable1 The specific values of some puree characteristics are shown at Table2

Figures 1-3 show the viscosity vs. shear rate plots for the purees containing 80%, 60% and 40% of solids from the original FVW mixture at different temperatures. First, it is clear that all mixtures show non-Newtonian behavior. The curves are typical of rheofluidizing fluids, as previously reported. As observed, the rheological curves are very different with regard to the solids contents, independent from the temperature. Curves are similar to those reported by Maceiras et al. [18] for raspberry, peach, prune and strawberry purees. It is important to remark that in this work, the shear rate range was 0.667 - 3.33 s^{-1}, while in that work, the range was wider (17.8 - 445 s^{-1}). The effect of temperature seems to be more pronounced for 80% and 60% solids samples.

Table 1: Composition of the fruits and vegetables purees and the mixed fruit/vegetables puree

Temp. °C	Solids content (%, w/v)				
	20	30	40	60	80
20	NC	40.403	106.49	440.67	1314.3
30	NC	38.643	106.49	377.83	1027.3
40	NC	36.073	82.629	286.09	844.28

*Adapted from Garcia-Peña et al. 2011.

Table 2: Characterization physical and chemical of fruits vegetables waste mixture

Parameter	Fruit vegetables waste
Soluble protein (µg/mL)	371.43 ± 2.24
Total carbohydrates (g/L)	57.65 ± 1.04
Chemical oxygen demand (mg/L)	144,133 ± 5292
Total organic carbon (% w/w)	2.84 ± 0.16
Lactate (g/L)	0.142 ± 0.014
Acetate (g/L)	ND
Ethanol (g/L)	ND
C% (w/w)	53.4 ± 9.5
N% (w/w)	2.13 ± 0.03
O% (w/w)	39.7 ± 8.61
P% (w/w)	0.18 ± 0.04
S% (w/w)	0.16 ± 0.03
Na% (w/w)	0.09 ± 0.01
Mg% (w/w)	0.09 ± 0.03
K% (w/w)	2.17 ± 0.02
Ca% (w/w)	0.12 ± 0.02
Fe (mg/L)	0.752
Ni (g/L)	ND
Density (g/mL)	0.919 ± 0.19
Total solids (g/L)	45.5 ± 0.99
pH	5.27

ND-undetectable, under detection limit.

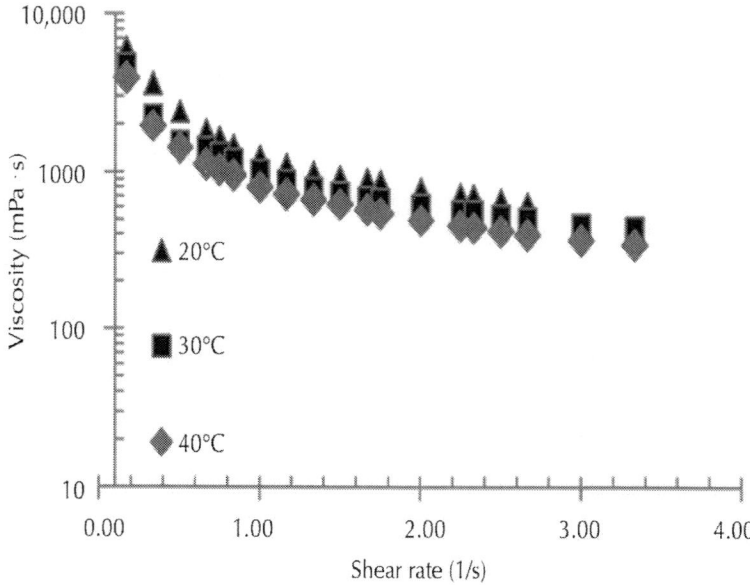

Figure 1: Viscosity vs. shear rate for the 80% solids puree.

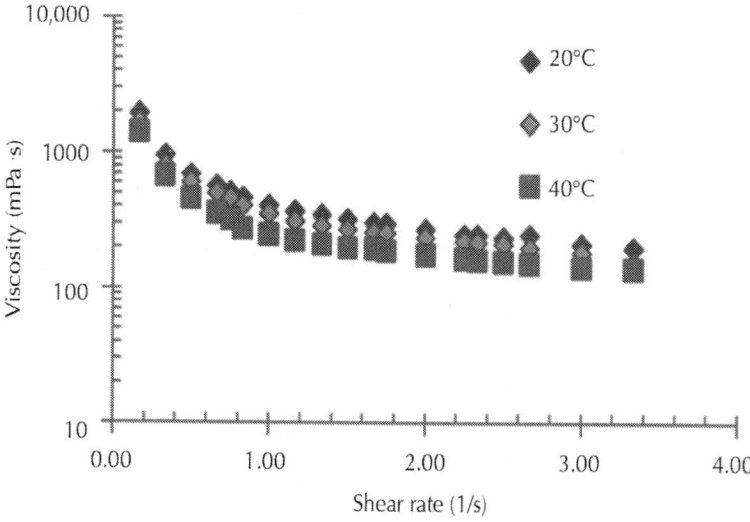

Figure 2: Viscosity vs. shear rate for the 60% solids puree.

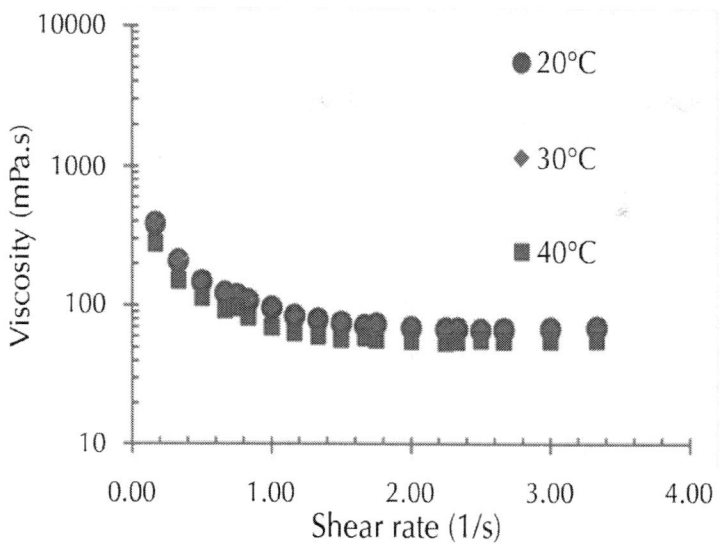

Figure 3: Viscosity vs. shear rate for the 40% solids puree.

Rheograms are also similar those reported by Nindo et al. [16] for blueberry puree. Again, in that work a wider range of shear rates were applied (10 - 1000 s⁻¹). Temperatures on that study where of 40°C, 50°C and 60°C. Nevertheless the shape of the curves is very similar to results showed inFigures 1-3.

Rheological data were fitted to the power law model, (see Figure 4 for the 80% puree). Correlation factors R^2 for the different mixtures were as follows. 0.991 - 0.995 for 80%, 0.961 - 0.986 for 60% and, 0.890 - 0.925, for 40%. In the case of the 30% puree, R^2 values were not acceptable (i.e., 0.41 - 0.65) and was observed that this flow is very near to the Newtonian behavior.

Tables 3 and 4 show the specific K and n values for the different mixtures at the three temperature values. Values for 30% puree were not calculated, since the low R^2 value obtained in the correlation.

As previously stated, the most used equation in order to model raw rheological data is the Ostwald de wale or power law model. Very few works have employed more than one rheological model, as in the case of El-Mansy [14] for papaya, mango and mango/papaya blends.

Most of the rheological curves were fitted reasonably using the power law or the IPC paste model, as shown by the correlation values obtained. Bingham and Casson models are applied when it is necessary to evaluate yield stress values ($_o$, $_{oA}$). Table 5 summarizes the K and n values for the 80%, 60%, 40% and 30% mixtures, together with results for other fruit or vegetable purees, reported in the literature. In the case where evaluations were carried out at different temperatures, the range of temperatures employed is also reported.

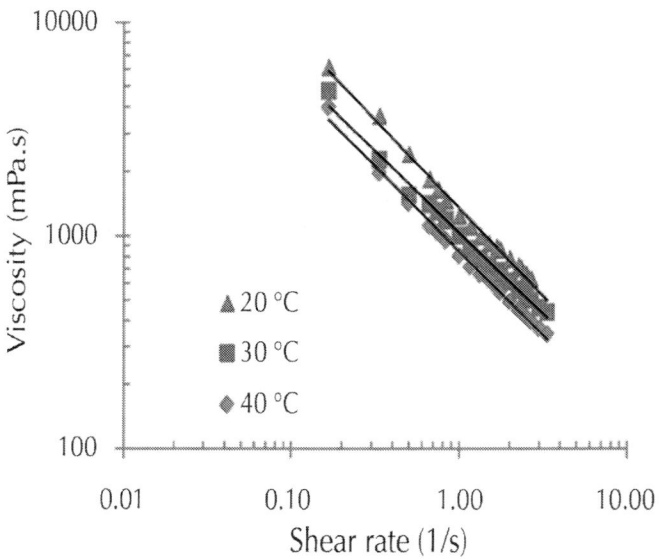

Figure 4: Viscosity vs. shear rate for the 30% solids puree. Solid lines represent the power law model.

Table 3: Consistency index (mPa·s) values for the fruits/vegetable puree at different concentrations

Temp.	Solids content (%, w/v)				
°C	20	30	40	60	80
20	NC	40.403	106.49	440.67	1314.3
30	NC	38.643	106.49	377.83	1027.3
40	NC	36.073	82.629	286.09	844.28

NC not calculated because the R^2 values were low.

Table 4: Index flow (undimensional) values for the fruits/vegetables puree at different concentrations

Temp. °C	Solids content (%, w/v)				
	20	30	40	60	80
20	NC	0.796	0.436	0.279	0.177
30	NC	0.749	0.436	0.275	0.237
40	NC	0.694	0.487	0.267	0.208

NC not calculated, since the R^2 values were low.

Effect of Temperature over Viscosity

Regarding the calculation of Ea, the activation energy, two different procedures have been reported for nonNewtonian fluids. The first one consist in correlate the K values with 1/T, and calculating the Ea values in mJ/mol. Andrade-Pizarro et al. [31] calculated Ea through the use of lnK and lnn. On the other hand, Ea can be calculated by fixing a shear rate in s^{-1}, so an apparent viscosity can be calculated. El-Mansy et al. [14] fixed the calculations at 10 rpm, arbitrarily.

For this work, it was decided to fix the shear rate at one value which could be found during the bio-hydrogen production in the 22 L tank. Fermentations have been carried out at 150 rpm, employing a pair of Rushton turbines. In order to calculate the shear rate, the Metzner and Otto correlation[32] was employed:

$$\gamma_{ave} = Ks \, N$$

(5)

where:

γ_{ave} is the average shear rate near the impeller in s^{-1};

Ks is a dimensionless physical constant (11.4 for Rushton turbines);

N is the agitation speed in s^{-1}.

By this procedure, Ea values were calculated for the 80%, 60% and 40% mixtures. Values of 15.98, 14.89 and 20.96 kJ/mol were found. They are showed on Table 5, with comparison purposes. The Ea value for the 20% - 30% mixtures were not calculated, since apparent viscosity was not well correlated by power law.

Figures 5 and 6 show the values of K and n, for every temperature (20°C - 40°C) and for every mixture concentration (30% - 80%). Regarding the consistency index, the value of temperature affects K values specially for. On the other hand, at low concentrated mixture (40%) the 60% and 80% mixtures, and the relationship is linear and inverse: the higher the temperature, the lower the K value K seems to be independent on T. It is yet possible to establish very simple equations to predict K as a function of T with fairly good R^2 values.

Table 5: Rheological characteristic of some fruits, vegetables and mixtures, K and n from Ostwald de Wale's power-law model, except when indicated

Fruit-vegetable-seed or mixture	Solids Brix (%, w/w)	K (Pa·s^n)	n (-)	T (°C)	Ea kJ/mol (kcal/g mol)	Reference
Blueberry	10 - 25	0.7 - 7.2	0.64 - 0.49	25 - 60	10.7 - 21.7	Nindo et al. (2007)
Raspberry	-	3.33 - 6.41	0.307 - 0.322	20 - 40	-	Maceiras et al. (2006)
Strawberry	-	6.96 - 8.64	0.222 - 0.238	20 - 40	-	
Prune	-	5.07 - 9.33	0.260 - 0.305	20 - 40	-	
Peach	-	11.50 - 15.39	0.251 - 0.268	20 - 40	-	
Zapote	29.26	637 - 3800	0.027 - 0.133	10 - 65	23.95	Andrade et al. (2010)
Papaya	11.86	3.15 - 25.61	0.30 - 0.40	5 - 100	-	El-Mansy et al. (2005)
Papaya/Mango blend P/M: 50/50 (w/w)	11.86/16.20	3.74 - 4.98	0.08 - 0.12	5 - 100	2.068	
Ginger paste (Herschel-Bulkley equation)	-	29.3 - 269.8	0.515 - 0.663	25 - 65	46.2	Ahmed (2004)
Milled maize and soy in water	33.4	7.81 - 22.11	0.12 - 0.28	80 - 160	-	Fraiha et al. (2011)

Tomato (Carreau model)	30.61	$\dot\eta_o = 2.8 \times 10^6$ Pa·s	$I_o = 8 \times 10^{-5}$ s^{-1}	25	-	Sanchez et al. (2009)
Fruit/ vegetable mixture in water ¨Solids as a percentage of the original puree	(80)¨	0.844 - 1.314	0.177 - 0.237	20 - 60	15.98	This work
	(60)¨	0.286 - 0.440	0.267 - 0.279	20 - 60	14.89	
	(40)¨	0.082 - 0.106	0.436 - 0.487	20 - 60	20.96	
	(30)¨	0.036 - 0.0404	0.694 - 0.796	20 - 60	NC	

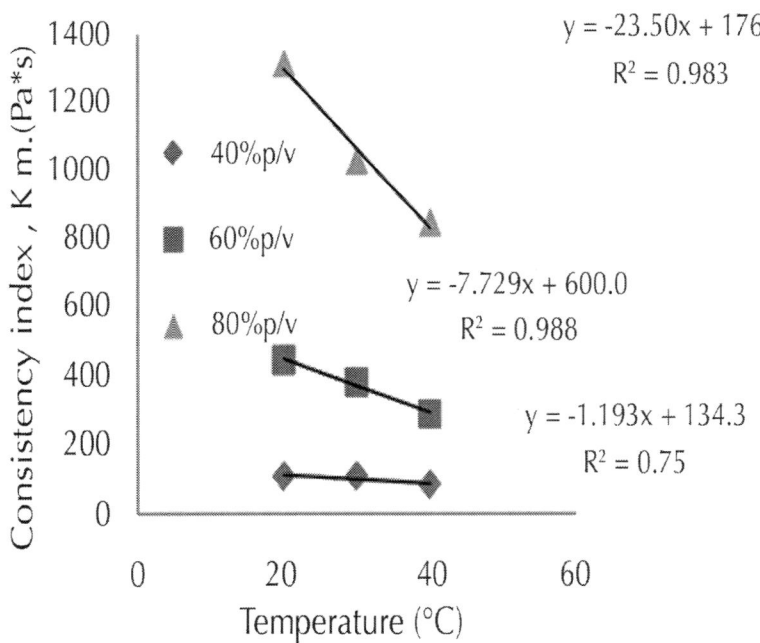

Figure 5: Consistency index values for different puree concentrations and temperatures.

Regarding the flow index, (Figure 6) it is evident that n values are affected by temperature for the 40% and 80% mixture. In the case of the mixture of 60%, seems that n value is independent on T. Again, it is possible to write very simple equations to predict n values as a function of T.

Effect of Time over Viscosity (Tyxothropy)

Raw data for 80%, 60%, 40% 30%, and 20% mixtures were obtained by duplicate. Viscosities calculated with the shear rate/shear stress values were quite similar for all the mixture concentrations. No apparent effect of time, i.e. thyrotrophic effects was noticed. As example show Figure 7, where upwards and downwards values of viscosities were calculated for the 80% mixture.

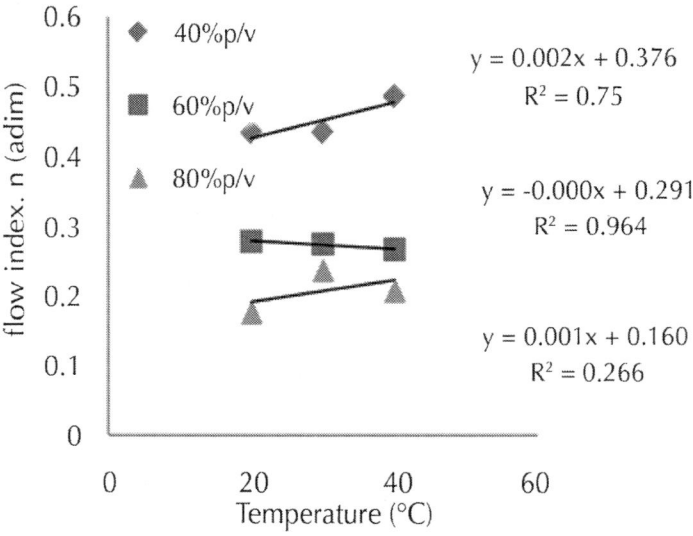

Figure 6: Effect of time over viscosity. 80% solids puree.

Simulation of the Puree's Rheological Behavior

The model employed to simulate the rheological behavior of purees was Carbopol (hydroxymethyl-cellulose), which rheological characteristics are dependent on both concentration and pH.

Figure 8 show the comparison of the rheograms for 80%, 60%, 40% and 30% mixture at 20°C, and the solid lines represent the Carbopol lines. As observed, Carbopol solutions have the capability of simulating fairly well the rheological behavior of the purees, in terms of the level of viscosity (K) and the pseudoplasticity of the fluids (n).

CONCLUSIONS

All mixtures show non-Newtonian behavior. The curves are typical rheofluidizing fluids, as previously reported. As observed, the rheological curves are very different with regard to the solids contents, independent from the temperature. Rheological data were fitted to the power law model. Correlation factors R^2 for the different mixtures were as follows: 0.991 - 0.995 for 80%, 0.961 - 0.986 for 60%, and 0.890 - 0.925, for 40%. In the case of the 30% puree, R^2 values were not acceptable. This flow is very near to the Newtonian behavior.

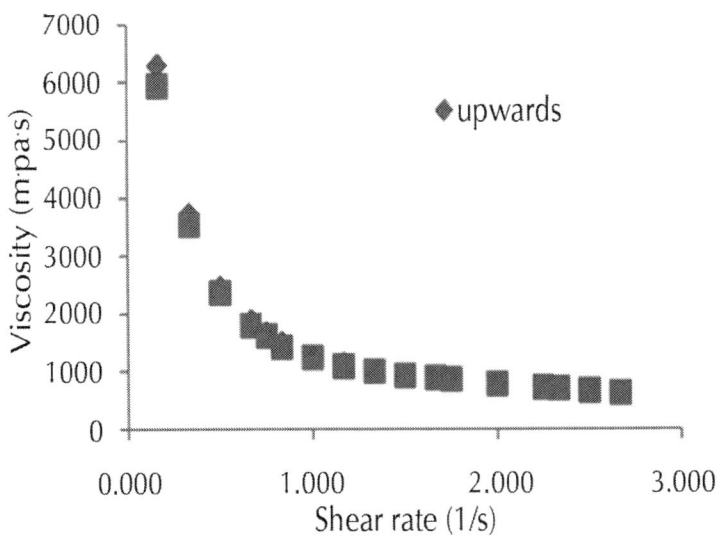

Figure 7: Comparison of 80%, 60% and 40% solids purees and Carbopol.

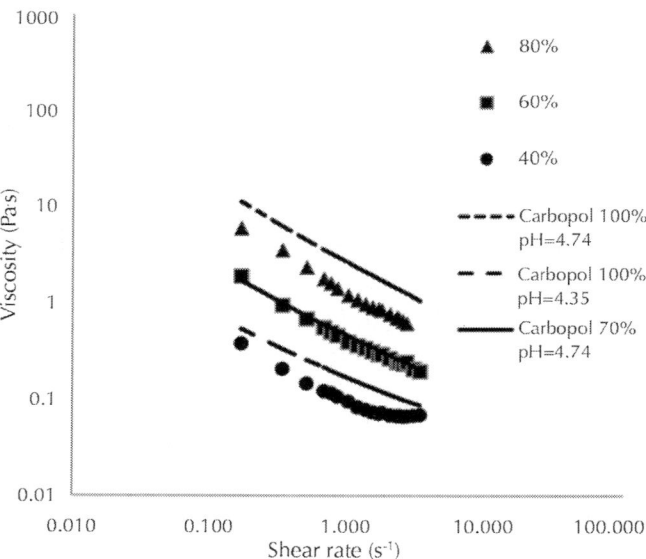

Figure 8: Viscosity versus shear rate for the 40% - 60% and 80% solids puree in comparison with the Carbopol solutions.

Ea values were calculated for the 80%, 60% and 40% mixtures. Values of 15.98, 14.89 and 20.96 kJ/mol were found. Finally, rheological behavior of FVW purees was well characterized by Carbopol solutions at given concentrations and pH values, as shown in Figure 6. Those measurements were carried out at 20°C.

ACKNOWLEDGEMENTS

Authors thank to PROMEP (Project Analisis integral de tratamiento de aguas residuales a traves de procesos biologicos), which supported this work. The participation of Marcos Martinez (UPIBI) in some rheological measurements is acknowledged and thanked.

REFERENCES

1.   B. E. Logan, S. E. Oh, I. Kim and S. Van Ginkel, "Biological Hydrogen Production Measured in Batch Anaerobic Respirometers,"

Environmental Science & Technology, Vol. 36, No. 11, 2002, pp. 2530-2535. http://dx.doi.org/10.1021/es015783i

2. J. Mata-Alvarez, F. Cecchi, P. Llabrés and P. Pavan, "Anaerobic Digestion of the Barcelona Central Food Market Organic Wastes: Plant Design and Feasibility Study," Bioresource Technology, Vol. 42, No. 1, 1992, pp. 33-42. http://dx.doi.org/10.1016/0960-8524(92)90085-C

3. S. N. Misi and C. F. Forster, "Semi-Continuous Anaerobic Co-Digestion of Agro-Waste," Environmental Technology, Vol. 23, No. 1, 2002, pp. 445-451.

4. H. Bouallagui, R. BenCheikh, L. Marouani and M. Hamdi, "Mesophilic Biogas Production from Fruit and Vegetable Waste in Tubular Digester," Bioresource Technology, Vol. 86, No. 1, 2003, pp. 85-90. http://dx.doi.org/10.1016/S0960-8524(02)00097-4

5. H. Bouallagui, Y. Touhami, R. BenCheikh and M. Hamdia, "Bioreactor Performance in Anaerobic Digestion of Fruit and Vegetable Wastes: Review," Process Biochemistry, Vol. 40, No. 3-4, 2005, pp. 989-995. http://dx.doi.org/10.1016/j.procbio.2004.03.007

6. E. I. Garcia-Peña, P. Parameswaran, D. W. Wang, M. Canul-Chan and R. Krajmalnik-Brown, "Anaerobic Digestion and Codigestion Process of Vegetable and Fruits Residues: Process and Microbial Ecology," Bioresource Technology, Vol. 102, No. 20, 2011, pp. 9447-9455. http://dx.doi.org/10.1016/j.biortech.2011.07.068

7. V. S. Mohan, G. Mohanakrishna, R. K. Goud and P. N. Sarma, "Acidogenic Fermentation of Vegetable Based Market to Harness Biohydrogen with Simultaneous Stabilization," Bioresource Technology, Vol. 100, No. 12, 2009, pp. 3061-3068.http://dx.doi.org/10.1016/j.biortech.2008.12.059

8. S. W. Van Ginkel, S. E. Oh and B. E. Logan, "Biohydrogen Gas Production from Food Processing and Domestic Wastewaters," International Journal of Hydrogen Energy, Vol. 30, No. 15, 2005, pp. 1535-1542. http://dx.doi.org/10.1016/j.ijhydene.2004.09.017

9. K. Vijayaraghavan, D. Ahmad and C. Soning, "Bio-Hydrogen Generation from Mixed Fruit Peel Waste Using Anaerobic Contact Filter," International Journal of Hydrogen Energy, Vol. 32, No. 18, 2007, pp. 4754-4760. http://dx.doi.org/10.1016/j.ijhydene.2007.07.001

10. P. Yang, R. Zhang, J. A. McGarveyc and J. R. Benemann, "Biohydrogen Production from Cheese Processing Wastewater by Anaerobic Fermentation Using Mixed Microbial Communities," International Journal of Hydrogen Energy, Vol. 32, No. 18, 2007, pp. 4761-4771. http://dx.doi.org/10.1016/j.ijhydene.2007.07.038

11. G. Mohanakrishna, R. Kannaiah Goud, S. Mohan and P. N. Sarma, "Enhancing Biohydrogen Production through Sewage Supplementation of Composite Vegetable Based Market Waste," International Journal of Hydrogen Energy, Vol. 35, No. 2, 2010, pp. 533-541. http://dx.doi.org/10.1016/j.ijhydene.2009.11.002

12. B. Ruggeri and T. Tommasi, "Efficiency and Efficacy of Pre-Treatment and Bioreaction for Bio-H_2 Energy Production from Organic Waste," International Journal of Hydrogen Energy, Vol. 37, No. 8, 2012, pp. 6491-6502. http://dx.doi.org/10.1016/j.ijhydene.2012.01.049

13. E. I. Garcia Peña, M. Canul-Chan, I. Chairez, E. Salgado and J. Aranda, "Continuous Bio-Hydrogen Production Based on the Evaluation of Kinetic Parameters of a Mixed Microbial Culture Using Fruit and Vegetable Wastes as Feedstock," Biomass Bioenergy, 2013.

14. H. A. El-Mansy, A. M. Sharoba, H. E. L. M. Bahlol and A. I. El-Desouky, "Rheological Properties of Mango and Papaya Nectar Blends," Annals of Agricultural Science, Moshtohor, Vol. 43, No. 2, 2005, pp. 665-686.

15. G. C. Antonio, F. R. Faria, C. Y. Takeiti and K. J. Park, "Rheological Behavior of Blueberry," Ciencia e Tecnologia de Alimentos, Vol. 29, No. 4, 2007, pp. 723-737.

16. C. L. Nindo, J. Tang, J. R. Powers and P. S. Tlakhar, "Rheological Properties of Blueberry Puree for Prossesing Application," LWT, Vol. 40, No. 2, 2007, pp. 292-299. http://dx.doi.org/10.1016/j.lwt.2005.10.003

17. R. D. Andrade, R. Torres, E. J. Montes and O. A. Perez, "Effect of Temperature on the Rheological Behavior of Zapote Pulp (Calocarpum sapota Merr)," Revista Tecnica de la Universidad de Zulia, Vol. 33, No. 2, 2010, pp. 138-144.

18. R. Maceiras, E. Alvarez and M. A. Candela, "Rheological Properties of Fruit Purees: Effect of Cooking," Journal of Food

Engineering, Vol. 80, No. 3, 2006, pp. 763-769.http://dx.doi.org/10.1016/j.jfoodeng.2006.06.028

19. M. C. Sanchez, C. Valencia, A. Ciruelos, A. Latorre and C. Gallegos, "Rheological Properties of Tomato Paste: Influence of the Addition of Tomato Slurry," Journal of Food Science, Vol. 68, 2006, pp. 551-554. http://dx.doi.org/10.1111/j.1365-2621.2003.tb05710.x

20. J. Ahmed, "Effect of Temperature on Rheological Characteristics of Ginger Paste," Emirates Journal of Agricultural Sciences, Vol. 16, No. 1, 2004, pp. 43-49.

21. J. Ahmed, Gangopadhaya and U. S. Shivhare, "Effect of Temperature on Rheological Characteristics of Green Chili Puree," Journal of Food Science and Technology, Vol. 36, 1999, pp. 352-354.

22. M. Fraiha, J. D. Biagi and A. C. de Oliveira, "Rheological Behavior of Corn and Soy Mix as Feed Ingredients," Ciencia e Tecnologia de Alimentos, Vol. 31, No. 1, 2011, pp. 129-134.http://dx.doi.org/10.1590/S0101-20612011000100018

23. Ansys Fluent Inc., "Fluent 13.0.," Lebanon, 2010.

24. B. Wu, "Advances in the CFD to Characterize Design and Optimize Bioenergy Systems," Computers and Electronics in Agriculture, Vol. 93, 2013, pp. 195-208.http://dx.doi.org/10.1016/j.compag.2012.05.008

25. P. Mavros, "Flow Visualization in Stirred Vessels. A Review of Experimental Techniques," Chemical Engineering Research and Design, Vol. 79, No. 2, 2001, pp. 113-127.http://dx.doi.org/10.1205/02638760151095926

26. E. L. Paul, V. A. Atiemo-Obeng and S. M. Kresta, "Handbook of Industrial Mixing," Science and Practice Wiley, New York, 2004.

27. T. Nagafune and Y. Hirata, "Measurement of Cavern Sizes and Shape in Agitated Yield Stress Aqueous Solutions with an Electrochemical Probe," 14th European Conference on Mixing Warszawa, 10-13 September 2012.

28. E. Galindo and A. W. Nienow, "Mixing of Highly Viscous Simulated Xanthan Fermentation Broths with the Lightnin A-315 Impeller," Biotechnology Progress, Vol. 8, No. 3, 1992, pp. 233-239. http://dx.doi.org/10.1021/bp00015a009

29. W. Kelly and B. Gigas, "Using CFD to Predict the Behavior of Power Law Fluids near Axial-Flow Impellers Operating in the Transitional Flow Regime," Chemical Engineering Science, Vol. 58, No. 10, 2003, pp. 2141-2152. http://dx.doi.org/10.1016/S0009-2509(03)00060-5

30. S. J. Curran, R. E. Hayes, A. Afacan, M. C. Williams and P. A. Tanguy, "Properties of Carbopol Solutions as Model for Yield-Stress Fluids," Journal of Food Science, Vol. 67, No. 1, 2002, pp. 176-180. http://dx.doi.org/10.1111/j.1365-2621.2002.tb11379.x

31. R. D. Andrade-Pizarro, R. Torres, E. J. Montes, O. A. Perez, C. E. Bustamante and B. B. Mora, "Effect of Temperature on the Rheological Behavior of Zapote Pulp (Calocarpum sapota Merr)," Revista Técnica de la Facultad de Ingeniería Universidad del Zulia, Vol. 33, No. 2, 2010, pp. 138-144.

32. B. Metzner and R. E. Otto, "Agitation of Non-Newtonian Fluids," AICHE Journal, Vol. 3, No. 1, 1975, pp. 3-11. http://dx.doi.org/10.1002/aic.690030103

Citations

CHAPTER 1

Byung-Doo Lee, Rajagopalan Thiruvengadathan, Sachidevi Puttaswamy, Brandon M Smith, Keshab Gangopadhyay, Shubhra Gangopadhyay, and Shramik Sengupta, Ultra-Rapid Elimination of Biofilms via The Combustion of a Nanoenergetic Coating, doi:10.1186/1472-6750-13-30.

CHAPTER 2

Alessandro Petruzzi and Francesco D'Auria, "Thermal-Hydraulic System Codes in Nulcear Reactor Safety and Qualification Procedures,"Science and Technology of Nuclear Installations, vol. 2008, Article ID 460795, 16 pages, 2008. doi:10.1155/2008/460795.

CHAPTER 3

O. Adeyefa and O. Oluwole, "Finite Element Modeling of Shop Built Spherical Pressure Vessels," Engineering, Vol. 5 No. 6, 2013, pp. 537-542. doi: 10.4236/eng.2013.56064.

CHAPTER 4

A. Valenzuela, J. Valenzuela and J. Parga, "Effect of Pretreatment of Sulfide Refractory Concentrate with Sodium Hypochlorite, Followed by Extraction of Gold by Pressure Cyanidation, on Gold Removal," Advances in Chemical Engineering and Science, Vol. 3 No. 3, 2013, pp. 171-177. doi: 10.4236/aces.2013.33021.

CHAPTER 5

Erler, J., Leistner, T. and Peuker, U. (2014) Application of a Particle Extraction Process at the Interface of Two Liquids in a Drop Column—Consideration of the Process Behavior and Kinetic Approach Advances in Chemical Engineering and Science, 4, 149-160. doi: 10.4236/aces.2014.42018.

CHAPTER 6

Parga, J. , Valenzuela, J. , Munive, G. , Vazquez, V. and Rodriguez, M. (2014) Thermodynamic Study for Arsenic Removal from Freshwater by Using Electrocoagulation Process. Advances in Chemical Engineering and Science, 4, 548-556. doi: 10.4236/aces.2014.44056.

CHAPTER 7

Oyedeko, K. and Susu, A. (2014) Design of a Simulator for Enhanced Oil Recovery Process Using a Nigerian Reservoir as a Case Study. Advances in Chemical Engineering and Science, 4, 430-453. doi:10.4236/aces.2014.44047.

CHAPTER 8

Yamauchi, S. , Saiki, S. , Ishibashi, K. , Nakagawa, A. and Hatakeyama, S. (2014) Low Pressure Chemical Vapor Deposition of Nb and F Co-Doped TiO2 Layer. Journal of Crystallization Process and Technology, 4, 79-88. doi: 10.4236/jcpt.2014.42011.

CHAPTER 9

Elboughdiri, N., Mahjoubi, A., Shawabkeh, A., Khasawneh, H. and Jamoussi, B. (2015) Optimization of the Degradation of Hydroquinone, Resorcinol and Catechol Using Response Surface Methodology. Advances in Chemical Engineering and Science, 5, 111-120. doi: 10.4236/aces.2015.52012.

CHAPTER 10

Isaac, I. and Nsi, E. (2015) Studies in Molecular Weight Determination of Cottonseed and Melon Seed Oils Based Biopolymers. Advances in Chemical Engineering and Science, 5, 43-50. doi: 10.4236/aces.2015.51005.

CHAPTER 11

Gomez-Romero, J., Garcia-Peña, I., Ramirez-Muñoz, J. and Torres, L. (2014) Rheological Characterization of a Mixed Fruit/Vegetable Puree Feedstock for Hydrogen Production by Dark Fermentation. Advances in Chemical Engineering and Science, 4, 81-88. doi: 10.4236/aces.2014.41011.

Index